About this book

This book is designed to help you get your best possible grade in your Mechanics 2 examination. The authors are Chief and Principal examiners and moderators, and have a good understanding of Edexcel's requirements.

Revise for Mechanics 2 covers the key topics that are tested in the Mechanics 2 exam paper. You can use this book to help you revise at the end of your course, or you can use it throughout your course alongside the course textbook, *Heinemann Modular Mathematics for Edexcel AS and A-level Mechanics 2* which provides complete coverage of the syllabus.

Helping you prepare for your exam

To help you prepare, each topic offers you:

- **Key points to remember** – summarise the mathematical ideas you need to know and be able to use.
- **Worked examples and examination questions** – help you understand and remember important methods, and show you how to set out your answers clearly.
- **Revision exercises** – help you practise using these important methods to solve problems. Exam-level questions are included so you can be sure you are reaching the right standard, and answers are given at the back of the book so you can assess your progress.
- **Test yourself questions** – help you see where you need extra revision and practice. If you do need extra help they show you where to look in the *Heinemann Modular Mathematics for Edexcel AS and A-level Mechanics 2* textbook.

Exam practice and advice on revising

Examination style paper – this paper at the end of the book provides a set of questions of examination standard. It gives you an opportunity to practise taking a complete exam before you meet the real thing. The answers are given at the back of the book.

How to revise – for advice on revising before the exam, read the How to revise section on the next page.

How to revise using this book

Making the best use of your revision time

The topics in this book have been arranged in a logical sequence so you can work your way through them from beginning to end. But **how** you work on them depends on how much time there is between now and your examination.

If you have plenty of time before the exam then you can **work through each topic in turn**, covering the key points and worked examples before doing the revision exercises and test yourself questions.

If you are short of time then you can **work through the Test yourself sections first**, to help you see which topics you need to do further work on.

However much time you have to revise, make sure you break your revision into short blocks of about 40 minutes, separated by five- or ten-minute breaks. Nobody can study effectively for hours without a break.

Using the Test yourself sections

Each Test yourself section provides a set of key questions. Try each question:

- If you can do it and get the correct answer then move on to the next topic. Come back to this topic later to consolidate your knowledge and understanding by working through the key points, worked examples and revision exercises.

- If you cannot do the question, or get an incorrect answer or part answer, then work through the key points, worked examples and revision exercises before trying the Test yourself questions again. If you need more help, the cross-references beside each test yourself question show you where to find relevant information in the *Heinemann Modular Maths for Edexcel AS and A-level Mechanics 2* textbook.

Reviewing the key points

Most of the key points are straightforward ideas that you can learn: try to understand each one. Imagine explaining each idea to a friend in your own words, and say it out loud as you do so. This is a better way of making the ideas stick than just reading them silently from the page.

As you work through the book, remember to go back over key points from earlier topics at least once a week. This will help you to remember them in the exam.

HEINEMANN MODULAR MATHEMATICS
for
EDEXCEL AS AND A-LEVEL
Revise for
Mechanics 2

John Hebborn   J~ ~ing

1A
1B
2
3
4
5

Heinemann

Edexcel
Success through qualifications

204 356

Heinemann Educational Publishers,
Halley Court, Jordan Hill, Oxford, OX2 8EJ
Part of Harcourt Education

Heinemann is the registered trademark of Harcourt Education Limited

First published 2001

05 04 03
10 9 8 7 6 5 4

ISBN 0 435 51114 9

Cover design by Gecko Limited

Original design by Geoffrey Wadsley; additional design work by Jim Turner

Typeset and illustrated by Tech-Set Limited, Gateshead, Tyne and Wear

Printed in the UK by Scotprint

Acknowledgements:

The publisher's and authors' thanks are due to Edexcel for permission to
reproduce questions from past examination papers. These are marked with an [E].

The answers have been provided by the authors and are not the responsibility of
the examining board.

Projectiles

Key points to remember

1 The horizontal speed of a projectile is unchanged throughout the motion.

2 The vertical motion has an acceleration of magnitude $g = 9.8 \, \text{m s}^{-2}$ vertically downwards.

3 The uniform acceleration equations can be used:

$$v = u + at$$

$$s = \left(\frac{u + v}{2}\right)t$$

$$v^2 = u^2 + 2as$$

$$s = ut + \tfrac{1}{2}at^2$$

4 The horizontal distance travelled by a projectile is called the **range** of the projectile.

5 To solve problems consider the horizontal and vertical motions separately.

For an initial velocity of magnitude u at an angle α above the horizontal:

    initial vertical velocity $= u \sin \alpha$ upwards
    constant horizontal speed $= u \cos \alpha$

Worked examination question 1 [E]

A ball is projected horizontally with speed $7 \, \text{m s}^{-1}$ from the top of a vertical tower which is 90 m high and which stands on a horizontal plane. Find how far from the bottom of the tower the ball strikes the plane.

Answer

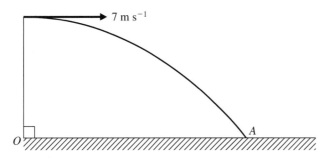

The horizontal distance OA is given by $7 \times$ time of flight.

To find the time of flight, consider the vertical motion.

Using $\quad s = ut + \frac{1}{2}at^2$

with $\quad s = 90\,\text{m}, u = 0\,\text{m s}^{-1}, a = 9.8\,\text{m s}^{-2}$

gives $\quad 90 = \frac{1}{2} \times 9.8t^2$

$$t^2 = \frac{2 \times 90}{9.8} = \frac{90}{4.9} = \frac{900}{49}$$

$$t = \sqrt{\left(\frac{900}{49}\right)} = \frac{30}{7}$$

So: $\quad OA = 7 \times \frac{30}{7}\,\text{m} = 30\,\text{m}$

Using **1**

Using **5**

Using **3**

Using **2**

Worked examination question 2 [E]

A projectile has an initial speed of $84\,\text{m s}^{-1}$ and rises to a maximum height of $40\,\text{m}$ above the level horizontal ground from which it was projected. Calculate:

(a) the angle of projection,
(b) the time of flight,
(c) the horizontal range.

Answer

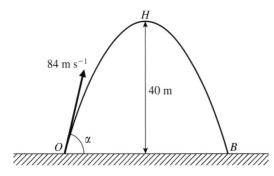

(a) Consider the vertical motion to the highest point, H.

Using $$v^2 = u^2 + 2as$$

Using **5**

Using **2**

with $s = 40\,\text{m}$, $u = 84 \sin \alpha\,\text{m s}^{-1}$, $v = 0\,\text{m s}^{-1}$, $g = -9.8\,\text{m s}^{-2}$

> $v = 0$ at the highest point. Acceleration due to gravity is always vertically downwards.

gives: $\quad 0 = (84 \sin \alpha)^2 - 2 \times 9.8 \times 40$

$\qquad (84 \sin \alpha)^2 = 2 \times 9.8 \times 40$

$\qquad 84 \sin \alpha = \sqrt{(2 \times 9.8 \times 40)}$

$\qquad \sin \alpha = \frac{1}{84}\sqrt{(2 \times 9.8 \times 40)}$

$\qquad \alpha = 19.47° = 19.5°$ (to 1 d.p.)

(b) Consider the vertical distance travelled by the projectile. Vertical distance travelled $= 0$ at the end of the flight.

> Using **5**

Using $\quad s = ut + \frac{1}{2}at^2$

> Using **3**

with $\quad s = 0\,\text{m}$, $a = -9.8\,\text{m s}^{-2}$, $u = 84 \sin \alpha\,\text{m s}^{-1}$

> Using **2** and **5**

gives: $\quad 0 = 84 \sin \alpha \times t - \frac{1}{2} \times 9.8 t^2$

$\qquad 0 = t(84 \sin \alpha - 4.9t)$

So: $\quad t = 0$ at start

or: $\quad t = \dfrac{84 \sin \alpha}{4.9} = 5.713$ at end.

> Use the value of α stored on your calculator for the most accurate answer.

So the time of flight is $5.71\,\text{s}$.

(c) Consider the horizontal motion.

> Using **5**

Using $\quad s = ut$ with $u = 84 \cos \alpha$

gives: $\quad s = 84 \cos \alpha \times 5.713\ldots$

$\qquad = 84 \cos 19.47 \times 5.713\ldots$

$\qquad = 452.5$

The horizontal range is $453\,\text{m}$.

> Using **4**

Worked examination question 3 [E]
A particle P, projected from a point O on horizontal ground, moves freely under gravity and hits the ground again at A. Referred to O as origin, OA as x-axis and the upward vertical at O as y-axis, the equation of the path of P is

$$y = x - \frac{x^2}{500},$$

where x and y are measured in metres.

(a) By finding $\dfrac{\mathrm{d}y}{\mathrm{d}x}$, show that P was projected from O at an angle of 45° to the horizontal.

(b) Find the distance OA and the greatest vertical height attained by P above OA.

(c) Find the speed of projection of P.

(d) Find, to the nearest second, the time taken by P to move from O to A.

Answer

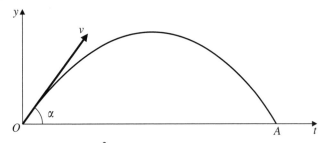

(a) Since $\quad y = x - \dfrac{x^2}{500}$

$$\frac{dy}{dx} = 1 - \frac{2x}{500}$$

So, when $x = 0$, $\dfrac{dy}{dx} = 1$.

> $\dfrac{dy}{dx}$ is the gradient of the tangent to the curve at the point (x, y).

At O, the velocity of P is at an angle of $\alpha = \arctan 1$ to the x-axis. Hence the angle of projection is $\alpha = 45°$ above the horizontal.

> The velocity is directed along the tangent to the curve.

(b) Along OA, $y = 0$.

So: $\quad 0 = x - \dfrac{x^2}{500} = x\left(1 - \dfrac{x}{500}\right)$

Hence: $x = 0$ (at O)

or: $\quad 1 - \dfrac{x}{500} = 0$

$$x = 500 \text{ (at } A)$$

The distance OA is $500\,\text{m}$.

The path of the projectile is symmetrical. So, at the greatest vertical height, $x = 250\,\text{m}$.

So: $\quad y = 250 + \dfrac{250^2}{500} = 375$

The greatest vertical height is $375\,\text{m}$.

(c) Consider the vertical motion.

> Using **5**

Using $s = ut + \frac{1}{2}at^2$

> Using **3**

with $\quad s = 0\,\text{m}$, $a = -9.8\,\text{m s}^{-2}$, $u = V\sin 45°\,\text{m s}^{-1}$

> Using **2** and **5**

gives $\quad 0 = V\sin 45° \times t - \frac{1}{2} \times 9.8t^2$

$$0 = t(V\sin 45° - 4.9t)$$

$$t = 0 \text{ (at } O) \quad \text{or} \quad V\sin 45° = 4.9t$$

So: $\quad t = \dfrac{V\sin 45°}{4.9}$ at A. $\hspace{2cm}$ (1)

Consider the horizontal motion.

At A, $s = 500$ m, so using distance $=$ speed $\times$ time

Using **5**

gives: $\qquad 500 = V\cos 45° \times \dfrac{V\sin 45°}{4.9}$

$$500 \times 4.9 = V^2 \sin 45° \times \cos 45°$$

$$500 \times 4.9 = V^2 \times \frac{1}{\sqrt{2}} \times \frac{1}{\sqrt{2}} = \frac{V^2}{2}$$

$$V^2 = 2 \times 500 \times 4.9$$

$$V^2 = 4900$$

$$V = 70$$

The speed of projection is $70 \, \text{m s}^{-1}$.

(d) From (1) above, t at A is $\dfrac{V\sin 45°}{4.9} = \dfrac{70\sin 45°}{4.9} = 10.1\ldots$

The time of flight is 10 s.

Worked examination question 4 [E]

A particle is projected with speed V at an angle of elevation α from a point A on a level plane. At time t, the particle passes through a point P which is vertically above a point Q on the plane. On passing through P the particle is moving in a direction which is at an angle of elevation θ. Given that the angle $PAQ = \beta$, express $\tan\beta$ and $\tan\theta$ in terms of V, α, g and t.

Hence, or otherwise, show that

$$2\tan\beta = \tan\alpha + \tan\theta.$$

A batsman hits a ball from a point on level ground so that the ball starts to move with speed $V\,\text{m s}^{-1}$ at an angle of elevation $\arctan\left(\frac{4}{5}\right)$. The ball is caught by a fielder at a height of 2 m above the ground when it is moving downwards at an angle $\arctan\left(\frac{2}{3}\right)$ to the horizontal. Calculate the distance between the batsman and the fielder.

Answer

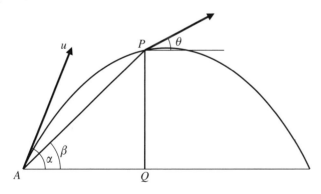

$$\tan\beta = \frac{PQ}{AQ} = \frac{\text{vertical distance to } P}{\text{horizontal distance to } P}$$

Consider the vertical motion to P.

Using $\quad s = ut + \frac{1}{2}at^2$

with $\quad u = V\sin\alpha,\ a = -g,\ t = t$

gives: $\quad PQ = (V\sin\alpha)t - \frac{1}{2}gt^2 \qquad (1)$

Consider the horizontal motion to P:

Using $s = ut$

with $\quad u = V\cos\alpha,\ t = t$

gives: $\quad AQ = (V\cos\alpha)t \qquad (2)$

Dividing $\dfrac{(1)}{(2)}$: $\quad \tan\beta = \dfrac{(V\sin\alpha)t - \frac{1}{2}gt^2}{(V\cos\alpha)t}$

$$= \frac{V\sin\alpha - \frac{1}{2}gt}{V\cos\alpha}$$

$$= \frac{V\sin\alpha}{V\cos\alpha} - \frac{\frac{1}{2}gt}{V\cos\alpha}$$

$$\tan\beta = \tan\alpha - \frac{gt}{2V\cos\alpha} \qquad (3)$$

$$\tan\theta = \frac{\text{vertical component of velocity at } P}{\text{horizontal component of velocity at } P}$$

Consider the vertical motion at P.

Using $v = u + at$

with $\quad u = V\sin\alpha,\ a = -g,\ t = t$

gives: $\quad$ vertical component of velocity at $P = V\sin\alpha - gt$

Horizontal component of velocity at $P = V\cos\alpha$

So $\quad \tan\theta = \dfrac{V\sin\alpha - gt}{V\cos\alpha}$

$$= \frac{V\sin\alpha}{V\cos\alpha} - \frac{gt}{V\cos\alpha}$$

$$= \tan\alpha - \frac{gt}{V\cos\alpha} \qquad (4)$$

$$\tan\alpha + \tan\theta = \tan\alpha + \left(\tan\alpha - \frac{gt}{V\cos\alpha}\right)$$

$$= 2\tan\alpha - \frac{gt}{V\cos\alpha}$$

$$= 2\left(\tan\alpha - \frac{gt}{2V\cos\alpha}\right)$$

$$= 2\tan\beta$$

Using **5**

Using **3**

Using **2** and **5**

Using **5**

Using **1**

Using **5**

Cancel t since $t \neq 0$.

Using $\tan\alpha = \dfrac{\sin\alpha}{\cos\alpha}$

Using **3**

Using **2** and **5**

Using **1**

Using equation (4)

From equation (3)

For the batsman hitting the ball, $\tan \alpha = \frac{4}{5}$ and $\tan \theta = -\frac{2}{3}$.

The ball is caught at a height of $2\,\text{m}$.

So: $\quad PQ = 2\,\text{m}$ and $AQ = \dfrac{2}{\tan \beta}\,\text{m}.$

As $\quad 2 \tan \beta = \tan \alpha + \tan \theta$

this gives: $\quad AQ = \dfrac{2}{\frac{1}{2}(\tan \alpha + \tan \theta)}$

$$= \dfrac{4}{\tan \alpha + \tan \theta}$$

$$= \dfrac{4}{\frac{4}{5} - \frac{2}{3}}$$

$$= \dfrac{4}{\frac{2}{15}} = 4 \times \tfrac{15}{2} = 30$$

The distance between the batsman and the fielder is $30\,\text{m}$.

> θ is an angle of elevation. For a ball moving downwards θ is negative.

Worked examination question 5 [E]

A particle is projected from a point O and, some time later, passes through the point with coordinates (x, y), where Ox and Oy are cartesian axes with Ox horizontal and Oy vertically upwards. Given that the velocity of projection has a horizontal component u and a vertical component v, show that

$$2yu^2 - 2uvx + gx^2 = 0.$$

Given that the particle passes through the points with coordinates $(36a, 6a)$ and $(48a, 4a)$, show that the velocity of projection is $13\sqrt{\left(\dfrac{ag}{2}\right)}$ at an elevation $\arctan\left(\frac{5}{12}\right)$.

Answer

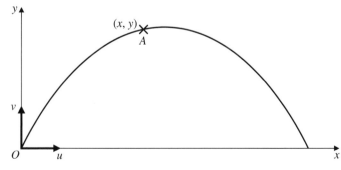

Let the point with coordinates (x, y) be A.

Consider the horizontal motion from O to A.

Since the horizontal speed is constant, $x = ut$. $\qquad$ (1) $\quad$ Using **1**

Consider the vertical motion from O to A.

Using $\quad s = ut + \frac{1}{2}at^2$

Using **3**

with $\quad s = y, u = v, a = -g$ and $t = t$

Using **2**

gives: $\quad y = vt - \frac{1}{2}gt^2$ $\hspace{4cm}$ (2)

From (1): $\quad t = \dfrac{x}{u}$

Substituting for t in (2) gives:

$$y = v\frac{x}{u} - \frac{1}{2}g\left(\frac{x}{u}\right)^2$$

$$y = \frac{vx}{u} - \frac{gx^2}{2u^2}$$

Multiplying through by $2u^2$ gives:

$$2yu^2 = 2uvx - gx^2$$

So: $\quad 2yu^2 - 2uvx + gx^2 = 0$ as required. $\hspace{2.5cm}$ (3)

The particle passes through the point with coordinate $(36a, 6a)$.

So: $\quad 2 \times 6au^2 - 2uv \times 36a + g(36a)^2 = 0$

> Substituting $x = 36a$ and $y = 6a$ in equation (3).

$$12au^2 - 72auv + 36^2a^2g = 0$$

$$au^2 - 6auv + 108a^2g = 0 \hspace{2cm} (4)$$

> Dividing by 12

The particle passes also through the point with coordinates $(48a, 4a)$.

So: $\quad 2 \times 4au^2 - 2uv \times 48a + g(48a)^2 = 0$

> Substituting $x = 48a$ and $y = 4a$ in equation (3).

$$8au^2 - 96auv + 48^2a^2g = 0$$

$$au^2 - 12auv + 288a^2g = 0 \hspace{2cm} (5)$$

> Dividing by 8

Equation (4) $\times$ 2: $\quad 2au^2 - 12auv + 216a^2g = 0 \hspace{1.5cm} (6)$

Equation (6) $-$ equation (5):

$$au^2 + 216a^2g - 288a^2g = 0$$

$$au^2 = 72a^2g$$

$$u^2 = 72ag$$

Substituting for u^2 in equation (4) gives:

$$72a^2g - 6a\sqrt{(72ag)}v + 108a^2g = 0$$

$$6a\sqrt{(72ag)}v = 180a^2g$$

$$v = \frac{180a^2g}{6a\sqrt{(72ag)}}$$

$$= \frac{180ag}{6 \times 6\sqrt{(2ag)}}$$

$$= 5\sqrt{\left(\frac{ag}{2}\right)}$$

Using
$\sqrt{72} = \sqrt{(36 \times 2)} = 6\sqrt{2}$

The components of the initial velocity are:

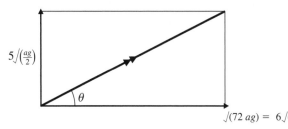

$5\sqrt{\left(\frac{ag}{2}\right)}$

θ

$\sqrt{(72\,ag)} = 6\sqrt{(2\,ag)}$

So the magnitude of the resultant velocity $= \sqrt{(u^2 + v^2)}$

$$= \sqrt{\left(72ag + 25\frac{ag}{2}\right)}$$

$$= \sqrt{\left(\frac{169ag}{2}\right)}$$

$$= 13\sqrt{\left(\frac{ag}{2}\right)}$$

The particle is projected at an angle of elevation θ where:

$$\tan\theta = \frac{5\sqrt{\left(\frac{ag}{2}\right)}}{6\sqrt{(2ag)}} = \frac{\frac{5}{2}\sqrt{(2ag)}}{6\sqrt{(2ag)}}$$

$$\tan\theta = \tfrac{5}{12}$$

Revision exercise 1A

1 A pebble is thrown horizontally with speed $6\,\mathrm{m\,s^{-1}}$ out to sea
   from a point on the top of a cliff which is at a height of $10\,\mathrm{m}$
   above sea level. The pebble moves freely under gravity. Find
   the horizontal distance travelled by the pebble before it hits
   the sea.   [E]

2 A particle A of mass $0.7\,\text{kg}$ is moving with speed $3\,\text{m}\,\text{s}^{-1}$ on a horizontal smooth table of height $0.9\,\text{m}$ above a horizontal floor. Another particle B, of mass $M\,\text{kg}$, is at rest on the edge of the table top. Particle A strikes particle B, and they coalesce into a single particle C. The particle C then falls from the table. From the point of leaving the table to the point of hitting the floor, the horizontal displacement of C is $0.6\,\text{m}$.

(a) Show that C takes $\frac{3}{7}\,\text{s}$ to fall to the floor.

(b) Find the value of M. [E]

3 An aircraft A is flying with constant speed $98\,\text{m}\,\text{s}^{-1}$ in a straight line and at a constant height of $1000\,\text{m}$ over the sea. At the instant when A is vertically above an anchored observation ship S, a bomb is released. The bomb falls freely under gravity and hits the sea at a point T.

(a) Explain why A is vertically above T at the instant when the bomb hits the sea.

Calculate:

(b) the time taken by the bomb to reach the sea from the instant of release from the aircraft,

(c) the distance of T from S,

(d) the vertical component of the velocity of the bomb at the instant when the bomb hits the sea. [E]

4 A ball is thrown from a point A on horizontal ground with speed $20\,\text{m}\,\text{s}^{-1}$ at an angle of elevation $56°$. The ball moves freely under gravity and after T seconds the ball hits the ground at a point B, where $AB = X$ metres. Giving your answers to 2 significant figures, find the value of

(a) T,

(b) X.

At time $2\,\text{s}$ after leaving A, the ball is at a height of H metres above the ground and has velocity $V\,\text{m}\,\text{s}^{-1}$. Find, to 2 significant figures,

(c) the value of H,

(d) the magnitude of V.

(e) Find the angle between V and the horizontal, to the nearest degree. [E]

5 A shell S is fired from a point O with speed $490\sqrt{2}\,\text{m s}^{-1}$ at an angle of elevation of $45°$. After t seconds the shell is at the point P whose coordinates are (x, y), referred to horizontal and upward vertical axes through O and in the plane of flight.

(a) Find an equation relating x and y.

(b) Calculate the horizontal range of S.

(c) Calculate the time taken by S to reach its greatest height above the level of O.

(d) State the minimum speed of S.   [E]

6

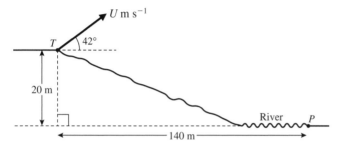

A golf ball is struck from a point T so as to clear a river. To clear the river, the golf ball must land at or beyond a point P on the far bank of the river, where P is at a horizontal displacement of 140 m from T and 20 m below the level of T, as shown in the diagram. The initial velocity of the ball has magnitude $U\,\text{m s}^{-1}$ and angle of elevation $42°$. The ball moves freely under gravity.

(a) Calculate, to 2 significant figures, the least value of U for which the ball clears the river.

(b) For $U = 40$, calculate, to 2 significant figures, the magnitude of the velocity of the ball 2 seconds after it has been struck.   [E]

7 A particle P is projected with speed $49\,\text{m s}^{-1}$ at an angle of elevation of θ, where $\sin\theta = \frac{3}{7}$, from a point A which is at a height of 17.5 m above a horizontal plane. The particle moves freely under gravity and first strikes the horizontal plane at the point B as shown in the diagram.

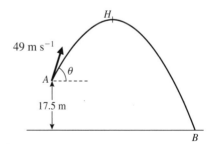

(a) Show that P reaches the highest point H of its path $2\frac{1}{7}\,\text{s}$ after leaving A.

(b) Find the vertical displacement between H and B and hence, or otherwise, deduce that P takes 5 s to move from A to B.

(c) Find the horizontal displacement between A and B, giving your answer to the nearest metre.

(d) Find the magnitude and the direction of the velocity of P at time 3 s after P leaves A.   [E]

8 A golf ball is driven from a point A with initial velocity of magnitude $42\,\text{m s}^{-1}$ at an angle of elevation $71.5°$. The ball moves freely under gravity and first strikes the ground at a point B on the same horizontal level as A.

(a) Calculate, in s to 1 decimal place, the time taken by the ball to reach B from A.

(b) Show that, to the nearest m, the distance AB is 108 m.

(c) Find, in m s^{-1} to 1 decimal place, the speed of the ball at an instant when the ball is moving at an angle of $25°$ to the horizontal.

A second ball is driven from A with initial velocity of magnitude $U\,\text{m s}^{-1}$ at an angle of elevation $\theta°$. This ball first strikes the ground at a point C on the same horizontal level as A. The distance AC is 120 m and the ball takes 4 s to reach C from A.

(d) Calculate to 1 decimal place, the values of U and θ.   [E]

Test yourself	**What to review**

If your answer is incorrect:

1 A cricket ball is hit from a point which is 1 m above horizontal ground. It is given a speed of $15\,\text{m s}^{-1}$ at an angle of elevation of $25°$.
Find:
(a) the time taken by the ball to reach the ground,
(b) the horizontal distance travelled by the ball.

Review Heinemann Book M2 pages 7–11

2 A particle is projected from a point on horizontal ground, with speed V at an angle α above the horizontal. The particle can be modelled as a projectile moving freely under gravity.
(a) Find, in terms of V, g and α, the range of the projectile.

A boy throws a stone with initial speed $u\,\mathrm{m\,s^{-1}}$ at an angle of elevation 40°. The stone strikes the ground at a point which is 24 m horizontally from the point of projection. In an initial model the stone is assumed to be a projectile moving freely under gravity which was thrown from a point on horizontal ground.
(b) Find, to 3 significant figures, the value of u.
(c) Find, in seconds, to 3 significant figures, the time of flight of the stone.
(d) State two physical features which could be included in order to make the model more realistic.

Review Heinemann Book M2 pages 7–11

3 A particle P is projected with speed $u\,\mathrm{m\,s^{-1}}$ at an angle of elevation θ, where $\sin\theta = \frac{3}{5}$, from a point on horizontal ground. The particle moves freely under gravity. It hits the ground at a point which is 480 m from its point of projection. Find:
(a) the value of u,
(b) the greatest height of P above the ground,
(c) the time of flight.
(d) State a physical factor which has been ignored in this model.

Review Heinemann Book M2 pages 7–11

4 A tennis player hits a ball at a point which is 2.1 m above horizontal ground, giving it a horizontal speed of $u\,\mathrm{m\,s^{-1}}$. The ball just clears the net which is 1 m high and 15 m horizontally from the point where the ball is hit.
(a) Find, to 3 significant figures, the value of u.
(b) Find, in $\mathrm{m\,s^{-1}}$ to 3 significant figures, the speed of the ball when it passes over the net.
(c) Find, to the nearest 0.1°, the direction of motion of the ball as it passes over the net.

Review Heinemann Book M2 pages 1–5

Test yourself answers

1 (a) 1.44 s **(b)** 19.5 m **2 (a)** $\dfrac{2V^2 \sin\alpha\cos\alpha}{g}$ **(b)** 15.5 **(c)** 2.03 s
  (d) two from: air resistance, spin, stone thrown above ground level, ground not horizontal
3 (a) 70 **(b)** 90 m **(c)** 8.57 s **(d)** air resistance, wind or spin
4 (a) 31.7 **(b)** 32.0 m s^{-1} **(c)** 8.3° below horizontal

Kinematics of a particle

Key points to remember

1 When the displacement of a particle P moving along a straight line is a function of time, the relationships between the displacement x, velocity v and acceleration a of P are shown by:

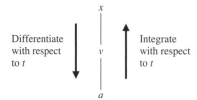

Differentiate with respect to t x Integrate with respect to t

v

a

2 If: $\qquad \mathbf{r} = x\mathbf{i} + y\mathbf{j}$
then: $\qquad \dot{\mathbf{r}} = \dot{x}\mathbf{i} + \dot{y}\mathbf{j}$
and $\qquad \ddot{\mathbf{r}} = \ddot{x}\mathbf{i} + \ddot{y}\mathbf{j}$

where the dot $\dot{}$ denotes differentiation with respect to t.

3 If the position vector $\mathbf{r}$ of a particle is given

then: $\qquad \mathbf{v} = \dot{\mathbf{r}} = \dfrac{\mathrm{d}\mathbf{r}}{\mathrm{d}t}$

and: $\qquad \mathbf{a} = \dot{\mathbf{v}} = \ddot{\mathbf{r}}$

$\qquad\qquad = \dfrac{\mathrm{d}\mathbf{v}}{\mathrm{d}t} = \dfrac{\mathrm{d}^2\mathbf{r}}{\mathrm{d}t^2}$

4 The speed of a particle is the magnitude of its velocity vector, $\mathbf{v}$, that is: $\qquad$ speed $= |\mathbf{v}|$

Worked examination question 1 [E]

A particle P moves in a straight line so that its velocity, $v\,\mathrm{m\,s^{-1}}$ at time t seconds where $t \geqslant 0$, is given by $v = 28 + t - 2t^2$.
Find:
(a) the time when P is instantaneously at rest,
(b) the speed of P at the instant when the acceleration of P is zero.

Given that P passes through the point O of the line when $t = 0$,
(c) find the distance of P from O when P is instantaneously at rest.

Answer

(a) When P is instantaneously at rest $v = 0$.

So: $\qquad\qquad 28 + t - 2t^2 = 0$

Factorising: $\quad (7 + 2t)(4 - t) = 0$

So: $\qquad\qquad t = -\frac{7}{2}$ (not possible since $t \geqslant 0$)

or: $\qquad\qquad t = 4$

(b) Acceleration $= \dfrac{\mathrm{d}v}{\mathrm{d}t} = 1 - 4t$

Using **1**

acceleration $= 0$ when $1 - 4t = 0$, so $t = \frac{1}{4}$

When $t = \frac{1}{4}$, $v = (28 + \frac{1}{4} - \frac{1}{8})\,\mathrm{m\,s^{-1}} = 28\frac{1}{8}\,\mathrm{m\,s^{-1}}$

(c) $x = \int (28 + t - 2t^2)\,\mathrm{d}t$

$\qquad = 28t + \dfrac{t^2}{2} - \dfrac{2t^3}{3} + c$

Using **1**, integrate v with respect to t to obtain x.

Do not forget the constant of integration.

As $x = 0$ when $t = 0$: $\quad c = 0$

So, when $t = 4$: $x = 112 + 8 - \frac{128}{3} = 77\frac{1}{3}$

P is at a distance $77\frac{1}{3}\,\mathrm{m}$ from O when instantaneously at rest.

Worked examination question 2 [E]

A particle P, moving in a straight line, passes through a fixed point O at time $t = 0$. At time t seconds the velocity, $v\,\mathrm{m\,s^{-1}}$ of P is given by

$$v = t^3 - \tfrac{7}{2}t^2 + kt + 4$$

where k is a constant. Given that $v = 2$ when $t = 2$, find, giving your answers to 3 significant figures where appropriate,
(a) the value of k,
(b) an expression for the acceleration of P at time t seconds,
(c) the maximum and minimum values of v,
(d) the distance of P from O when
$\qquad$ (i) $t = 1$, (ii) $t = 3$.
(e) Hence find, in $\mathrm{m\,s^{-1}}$ to 3 significant figures, the average speed of P in the interval $1 \leqslant t \leqslant 3$.

Answer

(a) Substituting $t = 2$ and $v = 2$ in $v = t^3 - \frac{7}{2}t^2 + kt + 4$ gives:

$$2 = 8 - 14 + 2k + 4$$

So: $k = 2$

(b) Acceleration $= \dfrac{\mathrm{d}v}{\mathrm{d}t} = 3t^2 - 7t + 2$

Using **1**

(c) Maximum and minimum values of v occur when $\dfrac{dv}{dt} = 0$.

So $\quad 0 = 3t^2 - 7t + 2 = (3t - 1)(t - 2)$

$\qquad t = \tfrac{1}{3}, \; t = 2$

Substituting $t = \tfrac{1}{3}$ in $v = t^3 - \dfrac{7}{2}t^2 + kt + 4$ gives:

$\qquad v = \left(\tfrac{1}{3}\right)^3 - \tfrac{7}{2}\left(\tfrac{1}{3}\right)^2 + 2\left(\tfrac{1}{3}\right) + 4 = 4.31 \quad$ (maximum value)

When $\quad t = 2, \; v = 2 \quad$ (minimum value)

> This information was given in the question.

(d) $x = \int \left(t^3 - \tfrac{7}{2}t^2 + kt + 4\right) dt$

$\qquad = \tfrac{1}{4}t^4 - \tfrac{7}{6}t^3 + t^2 + 4t + c$

> Using **1**, integrate v with respect to t to obtain x.

As $\quad s = 0$ when $t = 0$: $\quad c = 0$

(i) Substituting $t = 1$ gives $x = \left(\tfrac{1}{4} - \tfrac{7}{6} + 1 + 4\right)$ m $= 4.08$ m

(ii) Substituting $t = 3$ gives $x = \left(\tfrac{1}{4} \times 3^4 - \tfrac{7}{6} \times 3^3 + 9 + 12\right)$ m

$\qquad\qquad = 9.75$ m

(e) Average speed $= \dfrac{9.75 - 4.08}{3 - 1} = 2.83 \, \text{m s}^{-1}$

> Average speed
> $= \dfrac{\text{total distance}}{\text{total time}}$

Worked examination question 3 [E]

A particle P, of mass $5\,\text{kg}$ is acted on by a constant force $\mathbf{F}$. At time t seconds the position of the particle, $\mathbf{r}$ metres, is given by the equation

$$\mathbf{r} = (3t^2 - kt + 2)\mathbf{i} + (kt^2 + 4t - k)\mathbf{j},$$

where k is a positive constant.
(a) Find the acceleration of P in m s^{-2} in terms of k.

Given that the magnitude of $\mathbf{F}$ is $50\,\text{N}$, calculate
(b) the value of k.

The particle is moving in a direction parallel to $\mathbf{j}$ when $t = T$.
(c) Find the value of T.
(d) Hence find, to the nearest $0.1°$, the angle that the position vector makes with the direction of $\mathbf{i}$ when $t = T$.

Answer

(a) $\mathbf{r} = (3t^2 - kt + 2)\mathbf{i} + (kt^2 + 4t - k)\mathbf{j}$

$\qquad \mathbf{v} = \dfrac{d\mathbf{r}}{dt} = (6t - k)\mathbf{i} + (2kt + 4)\mathbf{j}$

> Using **3**

$\qquad \mathbf{a} = \dfrac{d\mathbf{v}}{dt} = 6\mathbf{i} + 2k\mathbf{j}$

(b) Using $\mathbf{F} = ma$ with $m = 5$ gives:

$$\mathbf{F} = 5(6\mathbf{i} + 2k\mathbf{j})$$

Hence: $|\mathbf{F}| = 5|6\mathbf{i} + 2k\mathbf{j}| = 5\sqrt{\{6^2 + (2k)^2\}}$

$$= 5\sqrt{(36 + 4k^2)}$$

$$= 10\sqrt{(9 + k^2)}$$

As $|\mathbf{F}| = 50\,\text{N}$ this gives: $50 = 10\sqrt{(9 + k^2)}$

$$5 = \sqrt{(9 + k^2)}$$

$$25 = 9 + k^2$$

$$k^2 = 16$$

$$k = \pm 4, \text{ but } k \text{ is positive, so } k = 4.$$

(c) From part **(a)**: $\quad \mathbf{v} = \dfrac{\mathrm{d}\mathbf{r}}{\mathrm{d}t} = (6t - k)\mathbf{i} + (2kt + 4)\mathbf{j}$

From part **(b)**, $k = 4$, so: $\quad \mathbf{v} = (6t - 4)\mathbf{i} + (8t + 4)\mathbf{j}$

When $t = T$, $\quad\quad\quad\quad \mathbf{v} = (6T - 4)\mathbf{i} + (8T + 4)\mathbf{j}$

The particle is moving in a direction parallel to $\mathbf{j}$, so the $\mathbf{i}$ component of $\mathbf{v}$ is zero.

Hence: $\quad 6T - 4 = 0, \; T = \frac{2}{3}$.

(d) When $t = \frac{2}{3}$, $\mathbf{r} = (3(\frac{2}{3})^2 - 4 \times \frac{2}{3} + 2)\mathbf{i} + (4(\frac{2}{3})^2 + 4(\frac{2}{3}) - 4)\mathbf{j}$

$$= \tfrac{2}{3}\mathbf{i} + \tfrac{4}{9}\mathbf{j}$$

$$\tan\theta = \dfrac{\frac{4}{9}}{\frac{2}{3}} = \tfrac{4}{9} \times \tfrac{3}{2} = \tfrac{2}{3}$$

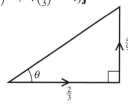

So: $\quad\quad \theta = 33.69\ldots^\circ = 33.7^\circ$

Worked examination question 4 [E]

An electric car is being tested on a large horizontal playground. At time T seconds, the position vector, $\mathbf{r}$ metres, of the car relative to a point O is given by

$$\mathbf{r} = \tfrac{9}{2}t^2\mathbf{i} + \tfrac{8}{5}t^{\frac{5}{2}}\mathbf{j}$$

At the instant when $t = 4$,

(a) show that the car is moving with velocity $(36\mathbf{i} + 32\mathbf{j})\,\text{m s}^{-1}$,

(b) find the magnitude of the acceleration of the car.

A cyclist is moving with constant velocity $17\mathbf{j}\,\text{m s}^{-1}$. At the instant when $t = 4$, calculate

(c) the velocity of the car relative to the cyclist,

(d) the speed of the car relative to the cyclist,

(e) the acute angle between this relative velocity and the constant velocity of the cyclist.

Answer

(a) $\mathbf{r} = \frac{9}{2} t^2 \mathbf{i} + \frac{8}{5} t^{\frac{5}{2}} \mathbf{j}$

$\mathbf{v} = \dfrac{d\mathbf{r}}{dt} = 9t\mathbf{i} + 4t^{\frac{3}{2}} \mathbf{j}$

Using **3**

Substituting $t = 4$ gives: $\mathbf{v} = 36\mathbf{i} + 32\mathbf{j}$
So the velocity of the car is $(36\mathbf{i} + 32\mathbf{j})\,\mathrm{m\,s^{-1}}$.

(b) $\mathbf{a} = \dfrac{d\mathbf{v}}{dt} = 9\mathbf{i} + 6t^{\frac{1}{2}} \mathbf{j}$

Using **3**

Substituting $t = 4$ gives: $\mathbf{a} = 9\mathbf{i} + 12\mathbf{j}$

So: $|\mathbf{a}| = \sqrt{(9^2 + 12^2)} = 15\,\mathrm{m\,s^{-2}}$

(c) $\mathbf{v}_{\text{car}} - \mathbf{v}_{\text{cyclist}} = 36\mathbf{i} + 32\mathbf{j} - 17\mathbf{j} = 36\mathbf{i} + 15\mathbf{j}$

Velocity of car relative to cyclist $= \mathbf{v}_{\text{car}} - \mathbf{v}_{\text{cyclist}}$ (see Book M1 Chapter 2).

(d) Speed of the car relative to the cyclist

$= |\mathbf{v}_{\text{car}} - \mathbf{v}_{\text{cyclist}}| = |36\mathbf{i} + 15\mathbf{j}|$

Using **4**

$= \sqrt{(36^2 + 15^2)} = 39\,\mathrm{m\,s^{-1}}$

(e) The cyclist is moving parallel to the vector $\mathbf{j}$.

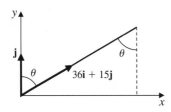

The required angle is θ, where $\tan\theta = \dfrac{36}{15}$,

$$\theta = 67.38\ldots^\circ = 67.4^\circ$$

Revision exercise 1B

1 At time t seconds, where $t \geqslant 0$, the acceleration of a particle P, moving in a straight line, is $(2t - 8)\,\mathrm{m\,s^{-2}}$. Given that the initial speed of P is $12\,\mathrm{m\,s^{-1}}$, find the times at which P comes instantaneously to rest. [E]

2 A particle P moves in a horizontal straight line. At time t seconds $(t \geqslant 0)$, the velocity, $v\,\mathrm{m\,s^{-1}}$, of P is given by $v = 12 - 4t$.
Find:
(a) the value of t when P is instantaneously at rest,
(b) the distance, in m, travelled by P between the time when $t = 0$ and the time when P is instantaneously at rest. [E]

3 A particle moves in a straight line such that at time t seconds its displacement, s metres, from a fixed point O of the line is given by

$$s = t^3 - 9t^2 + 24t - 7$$

Calculate

(a) the times when the particle is at rest,

(b) the greatest distance of the particle from O during the interval $0 \leqslant t < 4$. [E]

4 A rocket-propelled sledge moves along the x-axis and passes through the origin O with speed $4\,\mathrm{m\,s^{-1}}$ in the positive x-direction. At time t seconds after passing through O the acceleration of the sledge is $(6t + 2)\,\mathrm{m\,s^{-2}}$.

(a) Find the speed of the sledge when $t = 3$.

(b) Find the distance of the sledge from O at time t seconds.

(c) Hence find the distance covered by the sledge between the instants when $t = 2$ and $t = 4$.

(d) Sketch the speed–time curve for $0 \leqslant t \leqslant 4$. [E]

5 Particles P and Q move so that, at time t seconds, their position vectors, $\mathbf{r}_P$ metres and $\mathbf{r}_Q$ metres respectively relative to a fixed origin O, are given by

$$\mathbf{r}_P = (3t^2 + 5t)\mathbf{i} + 4t^2\mathbf{j}$$
$$\mathbf{r}_Q = 7t\mathbf{i} + 5t\mathbf{j}$$

(a) Show that the acceleration of P is constant, and find its magnitude.

(b) Find the velocity of P relative to Q at time t seconds.

(c) Find the time at which the direction of the velocity of P relative to Q is parallel to the vector $\mathbf{i} + \mathbf{j}$. [E]

6 A particle P moves so that, at time t seconds, its position vector $\mathbf{r}$ metres relative to a fixed origin O is given by

$$\mathbf{r} = t^2\mathbf{i} + (3t - t^2)\mathbf{j}, \qquad t \geqslant 0,$$

where the unit vectors $\mathbf{i}$ and $\mathbf{j}$ are directed due East and due North respectively.

(a) Find the magnitude of the velocity, in $\mathrm{m\,s^{-1}}$, of P when $t = 6$.

(b) Show that the acceleration of P is constant, and find the magnitude and bearing of this acceleration. [E]

7 A particle P moves along the x-axis passing through the origin O at time $t = 0$. At any subsequent time t seconds, P is moving with a velocity of magnitude $v \, \text{m s}^{-1}$ in the direction of x increasing where

$$v = 2t^3 + 2t + 3, \qquad t \geqslant 0.$$

(a) Find the acceleration of P when $t = 3$.

(b) Find the distance covered by P between $t = 0$ and $t = 4$.

A second particle Q leaves O when $t = 1$ with constant velocity of magnitude $10 \, \text{m s}^{-1}$ in the direction of the vector $3\mathbf{i} - 4\mathbf{j}$, where $\mathbf{i}$ and $\mathbf{j}$ are unit vectors parallel to Ox and Oy respectively.

Find, as a vector in terms of $\mathbf{i}$ and $\mathbf{j}$,

(c) the velocity of Q,

(d) the velocity of P relative to Q at the instant when $t = 1$.

Hence

(e) find the magnitude of the velocity of P relative to Q when $t = 1$,

(f) find the angle between the relative velocity and the vector $\mathbf{i}$ at this instant. [E]

Test yourself	**What to review**
	If your answer is incorrect:

1 A particle moves in a straight line such that its velocity $v \, \text{m s}^{-1}$ at time t seconds is given by

$$v = 3t^2 - t - 10, \qquad t \geqslant 0$$

Calculate
(a) the time, in s, when the particle is at rest,
(b) the acceleration, in m s^{-2}, when $t = 1$. [E]

Review Heinemann Book M2 pages 14–17

---

2 A particle P has mass $0.3 \, \text{kg}$. The position vector, $\mathbf{r}$ metres, of P relative to a fixed origin O, at time t seconds is given by

$$\mathbf{r} = 2t^3\mathbf{i} + (3t^2 - 5)\mathbf{j}, \qquad t \geqslant 0$$

(a) Find, in m s^{-1} to 3 significant figures, the speed of P when $t = 3$.
(b) Find, in N to 3 significant figures, the magnitude of the resultant force acting on P when $t = 3$.

Review Heinemann Book M2 pages 18–23

3 A particle P moves in a straight line. At time t seconds, the acceleration $a\,\mathrm{m\,s^{-2}}$ of P is given by

Review Heinemann Book M2 pages 14–17

$$a = 12t - 3t^2, \qquad t \geq 0$$

When $t = 0$, P is at rest at a fixed point O of the line.
(a) Calculate the velocity of P when $t = 3$.

P is next instantaneously at rest when $t = T$.
(b) Find the value of T.
(c) Find the average speed of P in the interval $0 \leq t \leq T$.

---

4 A particle P moves so that at time t seconds, $t \geq 0$, its position vector, $\mathbf{r}$ metres, relative to a fixed origin O is given by

Review Heinemann Book M2 pages 18–23

$$\mathbf{r} = (3t - t^3 + 2)\mathbf{i} + (t^2 + 2t)\mathbf{j}$$

(a) Find the velocity of P when $t = 4$.

The velocity of P is parallel to $(3\mathbf{i} - \mathbf{j})$ when $t = T$.
(b) Find the value of T.

Test yourself answers

1 (a) $2\,\mathrm{s}$ **(b)** $5\,\mathrm{m\,s^{-1}}$ **2 (a)** $56.9\,\mathrm{m\,s^{-1}}$ **(b)** $10.9\,\mathrm{N}$ **3 (a)** $27\,\mathrm{m\,s^{-1}}$ **(b)** $6\,\mathrm{s}$ **(c)** $18\,\mathrm{m\,s^{-1}}$
4 (a) $(-45\mathbf{i} + 10\mathbf{j})\,\mathrm{m\,s^{-1}}$ **(b)** $3\,\mathrm{s}$

Centres of mass

Key points to remember

1 The centre of mass of a lamina is the point at which the weight acts.

2 The weight of a **uniform** lamina is evenly distributed.

3 The centre of mass for a lamina or a discrete mass distribution must lie on any axis of symmetry.

4 The centre of mass of a set of n masses $m_1, m_2, \ldots, m_n$ at the points with coordinates $(x_1, y_1), (x_2, y_2), \ldots, (x_n, y_n)$ has coordinates $(\bar{x}, \bar{y})$ where:

$$\bar{x} = \frac{\sum m_i x_i}{\sum m_i} \quad \text{and} \quad \bar{y} = \frac{\sum m_i y_i}{\sum m_i}$$

5 Standard results:

Body	Centre of mass
uniform rod	mid-point of rod
uniform rectangular lamina	point of intersection of lines joining the mid-points of opposite sides
uniform circular disc	centre of circle
uniform triangular lamina	$\frac{2}{3}$ distance from any vertex to the mid-point of the opposite side
circular arc, radius r, angle at centre 2α	$\dfrac{r \sin \alpha}{\alpha}$ from centre
sector of circle, radius r, angle at centre 2α	$\dfrac{2r \sin \alpha}{3\alpha}$ from centre

> The first three of these results are obtained from symmetry considerations.

> The second three results are given in the formula book. Ensure you can find them quickly and use them efficiently in the examination.

6 For a lamina which is freely suspended and hangs in equilibrium, the centre of mass will be vertically below the point of suspension.

> **7** For a lamina which is balanced on an inclined plane, the line of action of the weight must fall within the side of the lamina that is in contact with the plane.

Worked examination question 1 [E]

Particles of mass $2M$, xM and yM are placed at points whose coordinates are (2, 5), (1, 3) and (3, 1) respectively. Given that the centre of mass of the three particles is at the point (2, 4), find the values of x and y.

Answer

	Separate masses			Total mass
Mass (kg)	$2M$	xM	yM	$(2 + x + y)M$
x-coordinate	2	1	3	$\bar{x} = 2$
y-coordinate	5	3	1	$\bar{y} = 4$

Taking moments about the y-axis gives:

$$2M \times 2 + xM \times 1 + yM \times 3 = (2 + x + y)M \times 2$$
$$4 + x + 3y = 4 + 2x + 2y$$
$$y - x = 0 \quad \text{or} \quad y = x$$

Taking moments about the x-axis gives:

$$2M \times 5 + xM \times 3 + yM \times 1 = (2 + x + y)M \times 4$$
$$10 + 3x + y = 8 + 4x + 4y$$
$$x + 3y = 2$$

Substituting $y = x$ gives: $\quad 4y = 2$

$$y = \tfrac{1}{2} = x$$

Example 1

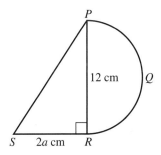

The diagram shows a frame made of uniform thin wire. The frame consists of a semicircular wire PQR together with straight wires PS, SR and PR. Triangle PSR is right-angled at R. The diameter of the semicircle is 12 cm and SR is of length $2a$ cm.

Given that the centre of mass of the framework lies on PR, show that $a^2 = 12$.

Answer

Centre of mass of a circular arc is $\dfrac{r\sin\alpha}{\alpha}$ from centre, where angle at centre $= 2\alpha$.

For the semicircle $\alpha = \frac{\pi}{2}$ and $r = 6\,\text{cm}$.

So: distance of centre of mass from the centre is $\dfrac{6\sin\frac{\pi}{2}}{\frac{\pi}{2}}\,\text{cm} = \dfrac{12}{\pi}\,\text{cm}$.

	semicircle	PR	PS	RS	whole frame
Ratio of masses	6π	12	$\sqrt{(12^2 + (2a)^2)}$	$2a$	M
Distance from PR	$-\dfrac{12}{\pi}$	0	a	a	0

Taking moments about PR gives:

$$-6\pi \times \frac{12}{\pi} + 12 \times 0 + \sqrt{(144 + 4a^2)} \times a + 2a \times a = M \times 0$$

$$a\sqrt{(144 + 4a^2)} = 72 - 2a^2$$

$$[a\sqrt{(144 + 4a^2)}]^2 = (72 - 2a^2)^2$$

$$a^2(144 + 4a^2) = 72^2 - 288a^2 + 4a^4$$

$$432a^2 = 72^2$$

$$a^2 = \frac{72^2}{432}$$

$$a^2 = 12$$

Using **5**

Using $\sin\dfrac{\pi}{2} = 1$

Find PS by using Pythagoras.

The mass of the whole frame is not needed as it will be multiplied by 0, since centre of mass of the frame is on PR.

The semicircle is on the opposite side of PR to PS and RS so the distance for the semicircle needs a negative sign.

Worked examination question 2 [E]

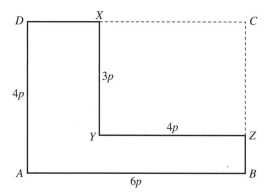

The diagram shows a uniform thin rectangular sheet of cardboard $ABCD$, which has $AB = 6p$ and $AD = 4p$. The rectangular sheet $XYZC$, where $YZ = 4p$ and $XY = 3p$, is cut away. Find the distance of the centre of mass of the L-shaped card $ABZYXD$ from

(a) AB,

(b) AD.

The L-shaped card is freely suspended from the corner D and hangs at rest in a vertical plane.

(c) Find the acute angle between DA and the downward vertical, to the nearest degree.

Answer

From the figure, $BZ = p$ and $DX = 2p$.

	$YZCX$	$ABZYXD$	$ABCD$
Ratio of masses	$12p^2$	$12p^2$	$24p^2$
Distance from AB	$\frac{5}{2}p$	$\bar{y}$	$2p$
Distance from AD	$4p$	$\bar{x}$	$3p$

> Distance of $YZCX$ from AB is length $BZ + \frac{1}{2}$ length ZC, and from AD is length $DX + \frac{1}{2}$ length XC.

(a) Taking moments about AB gives:
$$12p^2 \times \tfrac{5}{2}p + 12p^2\bar{y} = 24p^2 \times 2p$$
$$\tfrac{5}{2}p + \bar{y} = 4p$$
$$\bar{y} = \tfrac{3}{2}p$$

So the centre of mass is $\frac{3}{2}p$ from AB.

(b) Taking moments about AD gives:
$$12p^2 \times 4p + 12p^2\bar{x} = 24p^2 \times 3p$$
$$4p + \bar{x} = 6p$$
$$\bar{x} = 2p$$

So the centre of mass is $2p$ from AD.

(c) DG must be vertical.

Using **6**

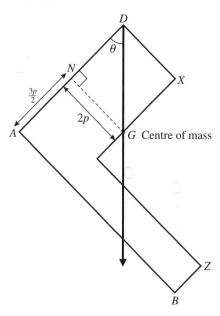

The required angle is θ.

$$\tan \theta = \frac{NG}{ND} = \frac{2p}{4p - \frac{3}{2}p} = \frac{2p}{\frac{5}{2}p} = \frac{4}{5}$$

$$\theta = 38.6\ldots^\circ = 39^\circ$$

Example 2

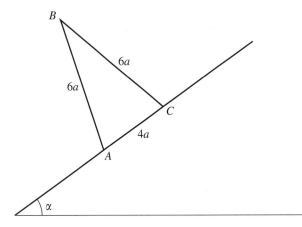

The uniform triangular lamina ABC has $AB = BC = 6a$ and $AC = 4a$. The diagram shows $\triangle ABC$ resting in a vertical plane on an inclined plane of angle α, with AC along a line of greatest slope. The plane is sufficiently rough to prevent slipping and the triangle is about to topple. Calculate, to the nearest 0.1°, the value of α.

Answer

By **7**, GA must be vertical and so the situation is as shown in the diagram.

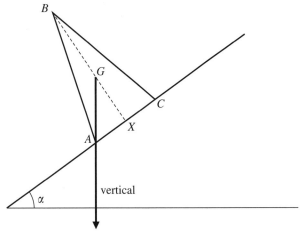

Let the perpendicular height of $\triangle ABC$ be BX and the centre of mass be G.

Then: $\quad BX = \sqrt{(36a^2 - 4a^2)}$ $\qquad\qquad$ By Pythagoras

$\qquad\quad = \sqrt{(32a^2)}$

$\qquad\quad = 4a\sqrt{2}$

The centre of mass of $\triangle ABC$ is $\frac{1}{3} \times 4a\sqrt{2}$ from AC, that is $GX = \dfrac{4a\sqrt{2}}{3}$.

$\angle GAX = (90° - \alpha)$ and $\angle AGX = \alpha$.

So $\tan \alpha = \dfrac{2a}{\dfrac{4a\sqrt{2}}{3}} = \dfrac{3}{2\sqrt{2}}$

$\qquad \alpha = 46.686\ldots°$

$\qquad \alpha = 46.7°$

Revision exercise 2

1 Particles of mass 2 kg, 3 kg and 12 kg are placed at the points with coordinates (2, 9), (3, 2) and (6, −2) respectively. Find the coordinates of the centre of mass of the three particles. [E]

2

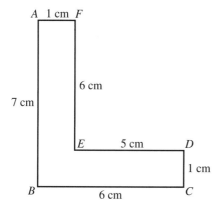

The diagram shows an outline of the letter L. The letter is 7 cm high and 6 cm wide with $CD = FA = 1$ cm. Find the distance of the centre of mass of this letter from AB,

(a) when the letter is a uniform lamina,

(b) when the letter is an outline made from uniform wire. [E]

3 The coordinates of the vertices O, A, B, C of a uniform rectangular metal plate $OABC$ are (0, 0), (6, 0), (6, 4), (0, 4) respectively. The square with vertices (2, 1), (3, 1), (3, 2), (2, 2) is removed from the plate.

(a) Show that the coordinates of the centre of mass of the remainder of the plate are $\left(\frac{139}{46}, \frac{93}{46}\right)$.

(b) Find, to the nearest degree, the angle which OA makes with the vertical when this remainder hangs freely from O. [E]

4

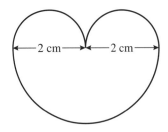

A badge is cut from a uniform thin sheet of metal. The badge is formed by joining the diameters of two semicircles, each of radius 1 cm, to the diameter of a semicircle of radius 2 cm, as shown in the diagram. The point of contact of the two smaller semicircles is O. Determine, in terms of π, the distance from O of the centre of mass of the badge.　　　　[E]

5

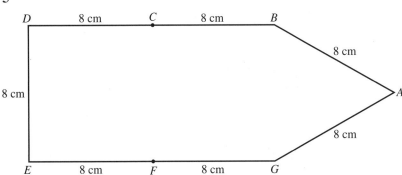

The diagram shows a sheet $ABCDEFG$ of uniform thickness. An open envelope is formed by folding the sheet along CF and by glueing CD to CB and FE to FG.

(a) Show that the distance of the centre of mass of the envelope from BG is $\dfrac{28}{8+\sqrt{3}}$. (You should ignore the mass of the glue.)

The envelope is suspended from the point B and hangs freely.

(b) Calculate, to the nearest $0.1°$, the angle between BA and the vertical.　　　　[E]

6 A uniform square plate $ABCD$ has mass $10M$ and the length of its side is $2l$. Particles of masses M, $2M$, $3M$ and $4M$ are attached at A, B, C and D respectively. Calculate, in terms of l, the distance of the centre of mass of the loaded plate from

(a) AB,

(b) BC.

The loaded plate is freely suspended from the vertex D and hangs in equilibrium.

(c) Calculate, to the nearest degree, the angle made by DA with the downward vertical. [E]

7

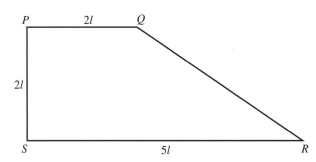

The diagram shows a uniform lamina $PQRS$ in which $PQ = PS = 2l$, $RS = 5l$ and $\angle QPS = \angle PSR = 90°$. Find the distance of the centre of mass of the lamina from

(a) PQ,

(b) PS.

The lamina stands with the edge PQ on a plane inclined to the horizontal at an angle θ. The lamina is in a vertical plane through a line of greatest slope of the plane. P is higher than Q. The lamina is on the point of toppling about Q.

(c) Find the value of $\tan \theta$.

| **Test yourself** | **What to review** |

If your answer is incorrect:

1

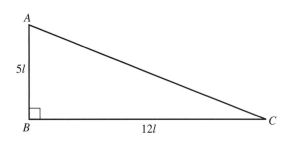

A thin uniform wire of length $30l$ is bent to form the sides of $\triangle ABC$, where $AB = 5l$, $BC = 12l$ and $\angle ABC = 90°$.
(a) Find the distance of the centre of mass of the wire from
(i) AB,
(ii) BC.

The wire is freely suspended from the mid-point of BC and hangs at rest.
(b) Find, to the nearest degree, the angle between BC and the vertical.

Review Heinemann Book M2 pages 45–47 and 50–52

2

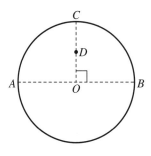

Review Heinemann Book M2 pages 39–43 and 50–52

A thin uniform circular disc has centre O, radius $5\,\text{cm}$ and mass $5m$. Particles of mass m and $3m$ are attached to the disc at the ends of the diameter AB.
(a) Explain briefly why the centre of mass of the loaded disc lies on AB.
(b) Find the distance of the centre of mass from O.

CO is a radius of the disc that is perpendicular to AB. D is the mid-point of CO. The disc is suspended by means of a smooth horizontal nail that passes through the disc at D. The disc is free to rotate in the vertical plane.
(c) Find, to the nearest degree, the angle between CO and the vertical.

3

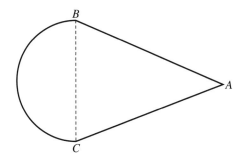

Review Heinemann Book M2 pages 41–43 and 50–52

The diagram shows a badge formed by joining a semicircle of radius a to an isosceles triangle ABC of base $2a$ and height $4a$. The badge is made from a uniform thin sheet of metal.
(a) Find the distance, in terms of a and π, of the centre of mass of the badge from the mid-point of the base of the triangle.

The badge is suspended from B and hangs freely under gravity.
(b) Find, to the nearest degree, the angle between BC and the vertical.

4

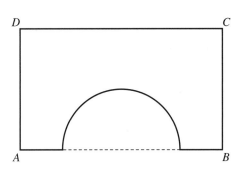

Review Heinemann Book M2 pages 43–44 and 52–53

The diagram shows the template made in order to produce the front face of a bridge for a model railway. $ABCD$ is a rectangle with $AB = 15$ cm and $BC = 8$ cm. A semicircle of diameter 10 cm and centre at the mid-point of AB is cut away.
(a) Find the distance of the centre of mass of the template from AB.

The template is placed upright on a rough inclined plane of angle α. AB lies along a line of greatest slope of the plane. Assuming that the plane is sufficiently rough to prevent sliding,
(b) find the maximum value of α for which the template can remain in equilibrium.

Test yourself answers

1 (a) (i) $5l$ (ii) $\dfrac{3l}{2}$ **(b)** $56°$

2 (a) AB is an axis of symmetry, or centres of mass of particles and disc are all on AB
(b) $1\frac{1}{9}$ cm **(c)** $24°$

3 (a) $\dfrac{28a}{3(\pi + 8)}$ **(b)** $40°$ **4 (a)** 4.91 cm **(b)** $56.8°$

Work, energy and power

3

Key points to remember

1 **Work**

For a force acting in the direction of the motion:

Work done = force × distance moved

For a force acting in a direction other than that of the motion:

Work done = component of force in the direction of motion × distance moved in the same direction

A force of 1 newton (N) does 1 joule (J) of work when moving a particle a distance of 1 metre.

2 **Energy**

The kinetic energy (K.E.) of a body of mass m moving with speed $v \, \text{m s}^{-1}$ is given by:

$$\text{K.E.} = \tfrac{1}{2}mv^2$$

For a mass in kg and velocity in m s^{-1} the K.E. is measured in joules (J). K.E. is never negative.

The potential energy (P.E.) of a body of mass m at a height h above a chosen fixed level is given by:

$$\text{P.E.} = mgh$$

Potential energy is also measured in joules. P.E. can be negative.

3 **Work–energy principle**

The work done on a particle is equal to its change of mechanical energy which is K.E. + P.E.

4 **Conservation of energy**

If the weight of a particle is the only force having a component in the direction of motion, then throughout the motion:

$$\text{K.E.} + \text{P.E.} = \text{constant}$$

---

5 **Power**
Power is the rate of doing work.
For a moving particle:

> Power = driving force × speed

Power is measured in watts, where 1 watt (W) is 1 joule per
second, or kilowatts, where 1 kilowatt (kW) is 1000 watts.

Worked examination question 1 [E]

At a mine an engine raises coal of mass 1200 kg from a vertical
depth of 400 m. The coal starts from rest and ends at rest and the
time required to complete this operation is 210 s. Calculate, in kW,
the average rate of working by the engine which raises this coal.

Answer

Initial K.E. = final K.E. = 0

> P.E. gained by the coal = $mgh = 1200 \times 9.8 \times 400$ J

So: work done by the engine = $1200 \times 9.8 \times 400$ J

Hence:

$$\text{average rate of working} = \frac{1200 \times 9.8 \times 400}{210} \text{ W}$$

$$= \frac{1200 \times 9.8 \times 400}{210 \times 1000} \text{ kW}$$

$$= 22.4 \text{ kW}$$

| At rest means $v = 0$, so K.E. = 0. |

| Using **2** |

| Using **3** |

| Average rate of working $= \dfrac{\text{total work done}}{\text{time taken}}$ |

Worked examination question 2 [E]

The magnitude of the resistance to the motion of a motor coach is
K newtons per tonne, where K is a constant. The motor coach has
mass $4\frac{1}{2}$ tonnes. When travelling on a straight horizontal road with
the engine working at 39.6 kW, the coach maintains a steady speed
of 40 m s^{-1}.
(a) Show that $K = 220$.

The motor coach ascends a straight road, which is inclined at
arcsin 0.3 to the horizontal, with the same power output and against
the same constant resisting forces.
(b) Find, in joules to 2 significant figures, the kinetic energy of the
motor coach when it is travelling at its maximum speed up the slope.

Answer

(a)

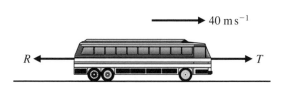

Let the tractive force be T and the resistance be R.
Resolving horizontally: $R = T$

Using **5**, power = force × speed, gives:

$$39.6 \times 10^3 = T \times 40$$

$$T = \frac{39.6 \times 10^3}{40}$$

Resistance = $4.5K$, so:

$$4.5K = \frac{39.6 \times 10^3}{40}$$

$$K = \frac{39.6 \times 10^3}{40 \times 4.5} = 220$$

> The tractive force is the force from the engine which pulls the coach forward.

> Steady speed means no acceleration.

(b)

$\sin \alpha = 0.3$

> The tractive force is not the same as in (a), so call it T_1.

$$R = 4.5K = 4.5 \times 220 \, \text{N} = 990 \, \text{N}$$

Using power = force × speed gives:

$$39.6 \times 10^3 = T_1 v \qquad (1)$$

Resolving parallel to the slope gives:

$$\begin{aligned} T_1 &= R + 4.5 \times 10^3 g \sin \alpha \\ &= 990 + 4.5 \times 10^3 \times 9.8 \times 0.3 \\ &= 14\,220 \end{aligned}$$

Using equation (1) gives: $39.6 \times 10^3 = 14\,220v$

Hence: $v = \dfrac{39.6 \times 10^3}{14\,220}$

> Using **5**

> There is no acceleration as the coach is moving at its maximum speed.

$$\begin{aligned} \text{K.E.} &= \tfrac{1}{2}mv^2 \\ &= \tfrac{1}{2} \times 4.5 \times 10^3 \times \left(\frac{39.6 \times 10^3}{14\,220} \right)^2 \, \text{J} \\ &= 17\,449 \, \text{J} \\ &= 17\,000 \, \text{J} \quad \text{(2 s.f.)} \end{aligned}$$

> Using **2**

Worked examination question 3 [E]

A pump raises 50 kg of water per second from a reservoir through a vertical height of 10 m. The water is discharged with speed $6\,\text{m s}^{-1}$. Find
(a) the potential energy, in J, gained by the water discharged each second,
(b) the kinetic energy, in J, gained by the water discharged each second.
(c) Hence find, in kW, the effective rate of working of the pump.

Answer

(a) P.E. gained by water discharged each second

$$mgh = 50 \times 9.8 \times 10\,\text{J} = 4900\,\text{J}$$

Using **2**

(b) K.E. gained by water discharged each second

$$\tfrac{1}{2}mv^2 = \tfrac{1}{2} \times 50 \times 6^2\,\text{J} = 900\,\text{J}$$

Using **2**

(c) Work done per second by the pump $= (4900 + 900)\,\text{J}$

Using **3**

$$\text{Rate of working} = 5800\,\text{J s}^{-1}$$
$$= 5800\,\text{W}$$
$$= 5.80\,\text{kW}$$

Example 1

A box of mass 20 kg slides down a line of greatest slope of a rough ramp that is inclined at angle θ to the horizontal where $\tan\theta = \frac{7}{24}$. The box starts from rest at point A and passes point B with speed $1.5\,\text{m s}^{-1}$. The distance from A to B is 2 m. By modelling the box as a particle, find
(a) the potential energy lost by the box in travelling from A to B,
(b) the kinetic energy gained by the box in travelling from A to B.

By using the work–energy principle,
(c) find the coefficient of friction between the box and the ramp.
(d) State a physical factor which has been ignored in this model.

Answer

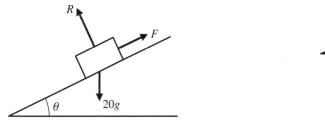

$$\cos\theta = \tfrac{24}{25}$$
$$\sin\theta = \tfrac{7}{25}$$

(a) P.E. lost by the box $= mgh$

Using **2**

$$= 20 \times 9.8 \times 2\sin\theta\,\text{J}$$
$$= 20 \times 9.8 \times 2 \times \tfrac{7}{25}\,\text{J}$$
$$= 109.76\,\text{J}$$

(b) Initial K.E. of box $= 0$

Final K.E. of box $= \frac{1}{2}mv^2$

$= \frac{1}{2} \times 20 \times 1.5^2 \, \text{J}$

$= 22.5 \, \text{J}$

Using **2**

So K.E. gained $= 22.5 \, \text{J}$

(c) By the work–energy principle:

Using **3**

Work done against friction $=$ energy lost

$= (109.76 - 22.5) \, \text{J}$

$= 87.26 \, \text{J}$

Work done against friction $= F \times 2 \, \text{J}$

The box has moved 2 m down the slope

So: $2F = 87.26$

$F = 43.63 \, \text{N}$

Resolving perpendicular to the ramp gives:

$R = 20g \cos \theta \, \text{N}$

$= 20 \times 9.8 \times \frac{24}{25} \, \text{N}$

$= 188.16 \, \text{N}$

Using $F = \mu R$ gives: $\quad 43.63 = 188.16\mu$

So: $\quad \mu = \dfrac{43.63}{188.16}$

$= 0.232$

(d) Air resistance has been ignored.

Example 2

A car of mass 1000 kg is moving along a straight horizontal road at a speed of $18 \, \text{m s}^{-1}$. The engine of the car is working at 20 kW and the total resistance to the motion of the car is 700 N.

(a) Find the acceleration of the car.

The car ascends a straight road which is inclined to the horizontal at an angle of 4°. The engine is now working at 30 kW and the non-gravitational resistance to the motion is still 700 N.

(b) Find the acceleration of the car at the instant when it is travelling at $20 \, \text{m s}^{-1}$.

Answer

(a)

$18 \, \text{m s}^{-1}$

700 N $\quad T$

Power $=$ force $\times$ speed
20 kW $= 20\,000$ W

$$\text{Tractive force} = \frac{20\,000}{18} \, \text{N}$$

Using $F = ma$ gives:

$$\frac{20\,000}{18} - 700 = 1000a$$

$$a = \frac{\dfrac{20\,000}{18} - 700}{1000} = 0.411$$

The acceleration is $0.411\,\mathrm{m\,s}^{-2}$.

(b)

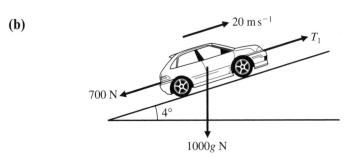

$$\text{Tractive force} = \frac{30\,000}{20}\,\mathrm{N} = 1500\,\mathrm{N}$$

Using $F = ma$ gives:

$$1500 - 700 - 1000g \sin 4° = 1000a$$

$$a = \frac{1500 - 700 - 1000g \sin 4°}{1000} = 0.116$$

The acceleration is $0.116\,\mathrm{m\,s}^{-2}$.

Revision exercise 3

1   A man hauls a bucket containing sand through a vertical
distance of 20 m. The mass of the bucket and the sand is
15 kg. The bucket starts from rest and finishes at rest.
(a) Calculate the work done by the man.
The man takes 45 s to raise the bucket.
(b) Calculate the average rate of working of the man.

2   A ladder of length 9 m is inclined at an angle of 70° to the
horizontal. A man of mass 75 kg slowly climbs the ladder from
the bottom to the top. Calculate the work done by the man.

3   A box of mass 20 kg is pulled in a straight line across a rough
horizontal floor. The coefficient of friction between the box
and the floor is 0.4. In an initial model, the applied force is
assumed to be constant and horizontal.
(a) Calculate the frictional force acting on the box.

The box is assumed to start from rest. When the box has moved a distance of $10 \, \text{m}$ the speed of the box is $6 \, \text{m s}^{-1}$.

(b) Calculate the magnitude of the applied force.

In a refined model the force is assumed to be applied at an angle of $10°$ above the horizontal. The magnitude of the force is unchanged. The box is jerked into motion and moves with an initial speed of $3 \, \text{m s}^{-1}$.

(c) Calculate the speed of the box when it has moved a distance of $10 \, \text{m}$.

4   The resistive forces opposing the motion of a car of mass $1200 \, \text{kg}$ are constant and of total magnitude $2000 \, \text{N}$.

(a) Find the power, in kW, which is required to keep the car moving along a straight level road at a constant speed of $20 \, \text{m s}^{-1}$.

When the car is moving along a straight level road at $20 \, \text{m s}^{-1}$ the rate of working of the engine of the car is suddenly increased and the initial acceleration of the car is $2 \, \text{m s}^{-2}$.

(b) Find the increase in the rate of working of the engine.

The car now comes to a hill of inclination $30°$ to the horizontal. If the rate of working is increased further to $100 \, \text{kW}$,

(c) find the constant speed with which the car can climb the hill. [E]

5   A train of total mass $3 \times 10^5 \, \text{kg}$ is moving at a constant speed of $10 \, \text{m s}^{-1}$ up a slope inclined at angle α to the horizontal, where $\sin \alpha = \frac{1}{120}$. The frictional resistance to the motion is of magnitude $7200 \, \text{N}$.

(a) Find, in kW to 2 significant figures, the rate at which the engine is working.

(b) Find, in m s^{-2} to 2 significant figures, the acceleration of the train at the instant at which it is travelling along a horizontal track with speed $10 \, \text{m s}^{-1}$, when the engine is working at $300 \, \text{kW}$ and the magnitude of the frictional resistance is unaltered. [E]

6 A car of mass $800 \, \text{kg}$ tows a caravan of mass $480 \, \text{kg}$ along a straight level road. The tow-bar connecting the car and the caravan is horizontal and of negligible mass. With the car's engine working at a rate of $24 \, \text{kW}$, the car and caravan are travelling at a constant speed of $25 \, \text{m s}^{-1}$.

(a) Calculate the magnitude of the total resistance, in N, to the motion of the car and the caravan.

The resistance to the motion of the car has magnitude $800 \, \lambda$ newtons and the resistance to the motion of the caravan has magnitude $480 \, \lambda$ newtons, where λ is a constant.

(b) Find the value of λ.

(c) Find the tension, in N, in the tow-bar. [E]

7 (a) Express $108 \, \text{km h}^{-1}$ in m s^{-1}.

A car of mass $900 \, \text{kg}$ is moving on a horizontal road at a constant speed of $108 \, \text{km h}^{-1}$, with its engine working at its maximum rate of $78 \, \text{kW}$.

(b) Calculate, in N, the total resistance acting on the car.

The car moves against this same non-gravitational resistance when moving up a hill inclined at $10°$ to the horizontal.

Given that the car is travelling up the hill at a constant speed and is working at the same rate of $78 \, \text{kW}$,

(c) calculate, in m s^{-1} to 2 s.f., the speed of the car.

The engine is switched off.

(d) Calculate, in m to 3 significant figures, the further distance that the car travels up the hill before coming to rest, assuming that the resistance remains unaltered. [E]

8 A car of mass $1000 \, \text{kg}$ is moving at a constant speed of $10 \, \text{m s}^{-1}$ up a straight road inclined at an angle α to the horizontal where $\sin \alpha = \frac{1}{15}$. The engine of the car is working at a constant rate of K kilowatts and the non-gravitational resistances to the motion of the car total R newtons.

(a) Find, in terms of R and g, an expression for K.

The car then moves down the same road at a constant speed of $30 \, \text{m s}^{-1}$. The engine is working at the same rate, K kilowatts, and the non-gravitational resistances now total $2R$ newtons. Find

(b) the value of R,

(c) the value of K.

Test yourself	What to review

If your answer is incorrect:

1 A box of mass 10 kg is being pulled up a line of greatest slope of a rough plane that is inclined at an angle of 15° to the horizontal. The coefficient of friction between the plane and the box is 0.25. At the instant when the box is moving at $10\,\text{m s}^{-1}$ the rope pulling the box breaks. Find the further distance moved by the box before coming to rest.

Review Heinemann Book M2 pages 68–71

2 A car of mass 900 kg is travelling along a straight horizontal road. The total resistance to the motion of the car is 600 N. When the car is moving with a speed of $25\,\text{m s}^{-1}$ the brakes are applied, producing a constant retarding force of 2200 N. Find the speed of the car when it has moved a distance of 50 m from the point where the brakes were first applied.

Review Heinemann Book M2 pages 68–71

3 A girl and her racing bike have a total mass of 65 kg. The girl is cycling along a straight horizontal road with constant speed $7\,\text{m s}^{-1}$ and she is working at a constant rate of 420 W.
(a) Calculate the magnitude of the force opposing her motion.

The girl now cycles down a straight road which is inclined at angle α to the horizontal. Her work rate is 420 W, the force opposing her motion is of magnitude 120 N and she is moving at a constant speed of $15\,\text{m s}^{-1}$.
(b) Calculate the value of α to the nearest degree. [E]

Review Heinemann Book M2 pages 74–76

4 A car of mass 700 kg is moving along a straight horizontal road against a constant resistive force of magnitude 450 N. The engine of the car is working at a rate of 8.76 kW.
(a) Find the acceleration of the car at an instant when its speed is $12\,\text{m s}^{-1}$.

The car now moves up a straight road inclined at $\arcsin(0.05)$ to the horizontal against the same resistive force of magnitude 450 N. The car moves at a constant speed $V\,\text{m s}^{-1}$ with the engine working at a rate of 12 kW.
(b) Find the value of V to 1 decimal place. [E]

Review Heinemann Book M2 pages 74–76

Test yourself answers

1 10.2 m   2 $17.7\,\text{m s}^{-1}$ 3 (a) 60 N (b) 8°   4 (a) $0.4\,\text{m s}^{-2}$ (b) 15.1

Collisions

Key points to remember

1 Momentum is a **vector**.

2 The **impulse** of a force = change in momentum produced

$$\mathbf{I} = m\mathbf{v} - m\mathbf{u}$$

3 **Conservation of linear momentum**

When two particles collide:

$$\begin{matrix} \text{total momentum} \\ \text{before the collision} \end{matrix} = \begin{matrix} \text{total momentum} \\ \text{after the collision} \end{matrix}$$

$$m_1 u_1 + m_2 u_2 = m_1 v_1 + m_2 v_2$$

$$\overset{m_1}{\bigcirc} \qquad \overset{m_2}{\bigcirc}$$

Before collision: $\longrightarrow u_1 \qquad \longrightarrow u_2$

After collision: $\longrightarrow v_1 \qquad \longrightarrow v_2$

4 **Newton's law of restitution**

$$\frac{\text{speed of separation of particles}}{\text{speed of approach of particles}} = e$$

where e is the coefficient of restitution between the particles.

5 Impact of a particle normally with a fixed surface.

$$\overset{m}{\bigcirc}$$

Before impact: $\longrightarrow u$

After impact: $v \longleftarrow$

$$(\text{speed of rebound}) = e\,(\text{speed of approach})$$

where e is the coefficient of restitution between the particle and the surface.

6 In general ($e \neq 1$) kinetic energy is **not** conserved in a collision.

Example 1

Two small smooth spheres A and B, each of mass 0.2 kg, are moving on the surface of a smooth horizontal table. Sphere A is travelling along a straight line with speed $18 \, \text{m s}^{-1}$ when it collides with sphere B, which is moving along the same straight line in the same direction with speed $9 \, \text{m s}^{-1}$. The coefficient of restitution between A and B is $\frac{1}{3}$.

(a) Find the speeds of A and B after the collision.
(b) Find the impulse that A exerts on B.

Answer

(a)
Before impact:

$\longrightarrow 18 \, \text{m s}^{-1}$ $\quad \longrightarrow 9 \, \text{m s}^{-1}$

A (0.2 kg) $\qquad B$ (0.2 kg)

After impact:

$\longrightarrow v_1 \, \text{m s}^{-1}$ $\quad \longrightarrow v_2 \, \text{m s}^{-1}$

A (0.2 kg) $\qquad B$ (0.2 kg)

Using the conservation of linear momentum

$$(0.2)18 + (0.2)9 = (0.2)v_1 + (0.2)v_2$$

So:
$$27 = v_1 + v_2 \qquad (1)$$

Using $\boxed{3}$

Using Newton's law of restitution

$$e = \tfrac{1}{3} = \frac{v_2 - v_1}{18 - 9}$$

Using $\boxed{4}$

So:
$$v_2 - v_1 = \tfrac{1}{3}(18 - 9)$$
$$v_2 - v_1 = 3 \qquad (2)$$

Adding equations (1) and (2) gives:

$$2v_2 = 30$$

So:
$$v_2 = 15$$

Substituting into (1):

$$v_1 = 27 - 15$$
$$= 12$$

After impact sphere A has speed $12 \, \text{m s}^{-1}$ and sphere B has speed $15 \, \text{m s}^{-1}$.

(b) Before impact momentum of B is $(0.2)9 = 1.8 \rightarrow$

After impact momentum of B is $(0.2)15 = 3 \rightarrow$

Using **2**, Impulse $=$ change in momentum

$$= (3 - 1.8) = 1.2\,\mathrm{N\,s}$$

> Only one particle should be considered when calculating an impulse.

Example 2

Two small smooth spheres A and B of equal radius and of mass $0.6\,\mathrm{kg}$ and $0.4\,\mathrm{kg}$ respectively, are moving on a smooth horizontal surface. Sphere A, moving with speed $12\,\mathrm{m\,s}^{-1}$, strikes directly sphere B which is initially moving in the opposite direction with speed $12\,\mathrm{m\,s}^{-1}$. After the impact sphere A is brought to rest.
(a) Find the speed of sphere B after the impact.
(b) Find e, the coefficient of restitution between the spheres.
(c) Find the loss of kinetic energy due to the impact.

Answer

Before impact:

$\longrightarrow 12\,\mathrm{m\,s}^{-1}$ $12\,\mathrm{m\,s}^{-1} \longleftarrow$

$A(0.6\,\mathrm{kg})$ $B(0.4\,\mathrm{kg})$

After impact:

$\longrightarrow 0\,\mathrm{m\,s}^{-1}$ $\longrightarrow v\,\mathrm{m\,s}^{-1}$

$A(0.6\,\mathrm{kg})$ $B(0.4\,\mathrm{kg})$

(a) Using the conservation of linear momentum

$$(0.6)12 - (0.4)12 = (0.4)v$$

> Using **3** and remember momentum is a vector **1**

So: $4v = 24$

$$v = 6$$

The speed of B after the impact is $6\,\mathrm{m\,s}^{-1}$.
(b) Newton's law of restitution gives:

> Using **4**

$$e = \frac{v}{12 + 12}$$

$$= \frac{v}{24}$$

and using the value of v found in **(a)**,

$$e = \tfrac{6}{24} = \tfrac{1}{4}$$

(c) The kinetic energy of the spheres **before impact** is:

$$\tfrac{1}{2}(0.6)(12)^2 + \tfrac{1}{2}(0.4)(12)^2 = 72$$

> Kinetic energy is a scalar.

The kinetic energy of the spheres **after impact** is

$$0 + \tfrac{1}{2}(0.4)(6)^2 = 7.2$$

So loss of kinetic energy due to impact is:

$$72 - 7.2 = 64.8\,\mathrm{J}$$

Example 3

A small smooth sphere A of mass $2m$ moving with speed V on a smooth horizontal table strikes directly another small smooth sphere B, of the same radius but of mass $3m$, which is at rest. The second sphere B then strikes a vertical wall at right angles and rebounds to meet A directly again. The coefficient of restitution between A and B is $\frac{1}{2}$ and that between B and the wall is $\frac{2}{3}$. Find the final speeds of the spheres.

Answer

1st impact

Before impact:

$\longrightarrow V$ $\longrightarrow 0$

$\bigcirc$ $\bigcirc$

$A\ (2m)$ $B\ (3m)$

After impact:

$\longrightarrow v_1$ $\longrightarrow v_2$

$\bigcirc$ $\bigcirc$

$A\ (2m)$ $B\ (3m)$

Conservation of linear momentum gives:

$$2mV = 2mv_1 + 3mv_2$$

So:

$$v_1 + \tfrac{3}{2}v_2 = V \qquad (1)$$

Using **3**

Newton's law of restitution gives:

$$e_1 = \tfrac{1}{2} = \frac{v_2 - v_1}{V}$$

So:

$$v_2 - v_1 = \tfrac{1}{2}V \qquad (2)$$

Using **4**

Adding equations (1) and (2):

$$\tfrac{5}{2}v_2 = \tfrac{3}{2}V$$

So:

$$v_2 = \tfrac{3}{5}V$$

Substituting in (1):

$$v_1 = V - \tfrac{3}{2}v_2$$

$$= V - \tfrac{3}{2}\left(\tfrac{3}{5}V\right) = \frac{V}{10}$$

Second impact

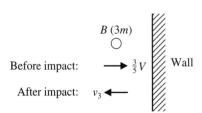

$$(\text{speed of rebound}) = e_2\,(\text{speed of approach})$$

Using **5**

So:

$$v_3 = \tfrac{2}{3}\left(\tfrac{3}{5}V\right) = \tfrac{2}{5}V$$

Third impact

Before impact:

$\xrightarrow{\frac{V}{10}}$ $\quad$ $\frac{2}{5}V \xleftarrow{}$

A (2m) $\qquad$ B (3m)

After impact:

$\xrightarrow{u_1}$ $\qquad$ $\xrightarrow{u_2}$

A (2m) $\qquad$ B (3m)

Conservation of linear momentum gives:

Using **3**

$$2m\left(\frac{V}{10}\right) - 3m\left(\frac{2V}{5}\right) = 2mu_1 + 3mu_2$$

So: $\qquad\qquad\qquad -V = 2u_1 + 3u_2 \qquad\qquad$ (3)

Newton's law of restitution gives:

Using **4**

$$e_1 = \tfrac{1}{2} = \frac{u_2 - u_1}{\frac{1}{10}V + \frac{2}{5}V}$$

So: $\qquad\qquad 2u_2 - 2u_1 = \frac{V}{2} \qquad\qquad$ (4)

Adding equations (3) and (4):

$$5u_2 = -\frac{V}{2}$$

So: $\qquad\qquad\qquad u_2 = -\frac{V}{10}$

Substituting into (3):

$$2u_1 = -V - 3u_2$$

$$= -V + \frac{3V}{10} = -\frac{7V}{10}$$

So: $\qquad\qquad\qquad u_1 = -\frac{7V}{20}$

The final speeds of A and B are $\frac{7V}{20}$ and $\frac{V}{10}$, both to the left:

$\xleftarrow{} \textcircled{A} \qquad \xleftarrow{} \textcircled{B}$

$\quad\frac{7V}{20} \qquad\qquad \frac{V}{10}$

Since the speed of A is greater than the speed of B there will not be any further collisions.

Example 4

A small rubber ball is dropped from a height of 18 m on to a horizontal floor. The coefficient of restitution between the ball and the floor is $\frac{2}{3}$. Find the height to which the ball rises on the rebound.

Answer

Using the constant acceleration formula

$$v^2 = u^2 + 2as$$

the speed v with which the ball reaches the floor is given by:

$$v^2 = 2g(18) = 36g$$

The speed V with which the ball rebounds is:

$$V = ev = \tfrac{2}{3}\sqrt{(36g)}$$

Using **5**

If h is the height reached after the rebound then using the constant acceleration formula again gives:

$$V^2 = 2gh$$

So:

$$h = \frac{V^2}{2g}$$

$$= \left(\frac{2}{3}\right)^2 \frac{36g}{2g}$$

$$= \frac{4 \times 36g}{9 \times 2g}$$

$$= 8$$

So on the rebound the ball rises to a height of 8 m.

Revision exercise 4

1 A sphere of mass $3m$ is moving with speed $2u$ when it collides with another sphere, of the same radius but of mass m, which is moving in the opposite direction with speed u. The coefficient of restitution between the spheres is $\frac{1}{3}$. Calculate:
   (a) the speed of each sphere immediately after impact,
   (b) the magnitude of the impulse received by each sphere on impact. [E]

2 A particle A, of mass $1.5\,\text{kg}$, is moving with speed $2\,\text{m s}^{-1}$ on a smooth horizontal table when it collides directly with another particle B, of mass $1\,\text{kg}$, which is moving with speed $1\,\text{m s}^{-1}$ in the same direction. After the collision A continues to move in the same direction with speed $1.4\,\text{m s}^{-1}$. Find:

  (a) the speed of B after impact,

  (b) the coefficient of restitution between A and B.

3 A particle A, of mass m, is moving on a smooth horizontal plane with speed v when it strikes directly a particle B, of mass $3m$, which is at rest on the plane. The coefficient of restitution between A and B is $\frac{1}{2}$ and after impact B moves with speed $3u$.

  (a) Find, in terms of u, the value of v and the speed of A after impact.

  Find also, in terms of m and u:

  (b) the magnitude of the impulse received by B,

  (c) the kinetic energy lost in the impact.   [E]

4 A smooth groove in the form of a circle of radius a is carved out of a horizontal table. Two small equal spheres, A and B, lie at rest in the groove at opposite ends of a diameter. At time $t = 0$ the sphere A is projected along the groove and the first collision occurs at time $t = T$. Given that e is the coefficient of restitution between the spheres, find the velocities of A and B after the first collision. Hence, or otherwise, show that the second collision takes place at time

$$t = T(2 + e)/e \qquad\qquad \text{[E]}$$

5 Three small smooth spheres A, B and C of equal size and of mass m, $2m$ and $2m$ respectively, lie at rest in a straight line on a smooth horizontal table, with B between A and C. The coefficient of restitution between any two spheres is e.

  Sphere A is projected directly towards B with speed u. Given that A rebounds from B with speed $\frac{1}{12}u$,

  (a) show that $e = \frac{5}{8}$.

  Sphere B then collides with sphere C.

  (b) Show that the speed of C after this collision is $\frac{169}{384}u$.

6 A small smooth ball falls from rest from a height of 3.2 m above a fixed smooth horizontal plane. It rebounds to a height of 1.25 m. Find the coefficient of restitution between the ball and the plane.

Test yourself	**What to review**
	If your answer is incorrect:
1 Three identical smooth spheres A, B, C each of mass 2 kg lie at rest on a smooth horizontal table. Sphere A is projected with speed $18 \, \text{m s}^{-1}$ to strike sphere B directly. Sphere B then strikes C directly. The coefficient of restitution between any two spheres is $\frac{1}{3}$. (a) Find the speeds of the spheres after these two collisions. (b) Find the total loss of kinetic energy due to these two collisions.	*Review Heinemann Book M2 pages 104, 114–117*
2 A small smooth ball, of mass 1.5 kg, is dropped from a height of 2 m on to a smooth horizontal plane and rebounds to a height of 1.2 m. (a) Find the value of e, the coefficient of restitution between the ball and the plane. (b) Find the loss of mechanical energy due to the impact.	*Review Heinemann Book M2 pages 104, 107–111*
3 A small smooth sphere A, of mass $3m$, is moving with speed u on a smooth horizontal table when it collides directly with another small smooth sphere B, of mass $5m$, which is at rest. The coefficient of restitution between A and B is e. (a) Find, in terms of e and u, the speeds of A and B immediately after the collision. Given that the speed of A after the collision is $\dfrac{u}{4}$ (b) find the value of e. (c) Find the magnitude of the impulse exerted by B on A.	*Review Heinemann Book M2 pages 95–104*

Test yourself answers

1 (a) $A \,(6 \, \text{m s}^{-1})$, $B \,(4 \, \text{m s}^{-1})$, $C \,(8 \, \text{m s}^{-1})$ (b) 208 J
2 (a) 0.775 (b) 11.8 J
3 (a) $\dfrac{u}{8}(3-5e)$, $\dfrac{3u}{8}(1+e)$ (b) $e = \frac{1}{5}$ (c) $\frac{9}{4}mu$

Statics of rigid bodies

5

Key points to remember

1 A rigid body is in **equilibrium** if:
(i) The vector sum of the forces acting is zero, that is the sum of the components of the forces in any given direction is zero.
(ii) The algebraic sum of the moments of the forces about any given point is zero.

2 Only in the case of limiting equilibrium, when motion is on the point of taking place, does the frictional force F have its maximum value μR.

3 To solve questions of this type:
either (i) resolve in two perpendicular directions and take moments about one point.
or (ii) take moments about two points and resolve in one direction.
Use $F = \mu R$, when applicable (see **2**).

Worked examination question 1 [E]

One end of a uniform ladder, of mass M and length $2l$, stands on rough horizontal ground, and the other end rests against a smooth vertical wall. The ladder rests in equilibrium in a vertical plane perpendicular to the wall and is inclined to the ground at an angle θ, where $\tan \theta = \frac{4}{3}$.

A man of mass $3M$ starts slowly to climb the ladder. Given that the coefficient of friction between the ladder and the ground is $\frac{1}{2}$, find, in terms of l, how far up the ladder the man can climb before the ladder begins to slip.

Answer

First draw a diagram to show all the given information with the man at the point P.

> At B there is only a normal reaction as the vertical wall is smooth.

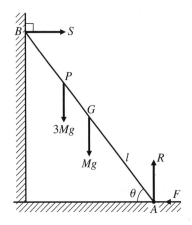

Let $AP = x$.

Resolving horizontally: $\rightarrow S - F = 0$ (1)

Resolving vertically: $\uparrow R - Mg - 3Mg = 0$ (2)

Taking moments about A:

$\curvearrowleft Mgl \cos\theta + 3Mgx \cos\theta - S\,2l \sin\theta = 0$ (3)   Using **3** (i)

By **2** as equilibrium is limiting:

$$F = \mu R \qquad\qquad (4) \qquad \text{Using } \mathbf{2}$$

From (2): $R = 4Mg$ (5)

From (1) and (4) and using $\mu = \frac{1}{2}$

$$S = F = \mu R = \tfrac{1}{2} R$$

and using (5): $S = \frac{1}{2}(4Mg) = 2Mg$ (6)

Using (6) in (3) gives:

$$Mgl \cos\theta + 3Mgx \cos\theta - (2Mg)2l \sin\theta = 0$$

or $l + 3x - 4l \tan\theta = 0$

Using $\tan\theta = \frac{4}{3}$

$$3x = 4l\left(\tfrac{4}{3}\right) - l$$
$$= l\left(\tfrac{16}{3} - 1\right)$$
$$= \tfrac{13}{3}l$$

So: $x = \frac{13}{9}l = 1\frac{4}{9}l$

The man may therefore climb a distance $1\frac{4}{9}l$ before the ladder begins to slip.

Worked examination question 2 [E]

A straight uniform rod $CDEF$, of length $6a$ and mass M, is suspended from a fixed point A by two light inextensible strings AD and AE, each of length $2a$, where D and E are the points of trisection of the rod. A particle of mass m is attached to the rod at a point distant x from the end C and the system hangs in equilibrium. Given that the tension in the string AD is twice the tension in the string AE, show that the rod is inclined to the horizontal at an angle θ where $\tan \theta = \dfrac{1}{3\sqrt{3}}$.

Hence find in terms of M, m, a and g
(a) the tensions in the strings,
(b) the distance x.

Answer

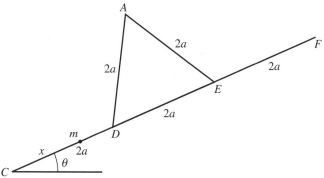

As D and E are points of trisection
$CD = DE = EF$
$\quad = \frac{1}{3}(6a) = 2a$

The diagram shows the given information.

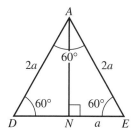

The triangle ADE is equilateral so that each angle is $60°$. If N is the mid-point of DE then $\angle ANE$ is a right angle.

Also $\quad AN = 2a \sin 60°$

$$= 2a \frac{\sqrt{3}}{2} = a\sqrt{3}.$$

Since AN is perpendicular to $CDEF$, which makes an angle θ with the horizontal, AN makes an angle θ with the vertical AQ.

We have then:

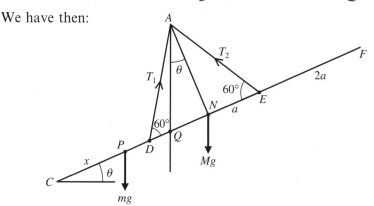

It is essential in a problem of this kind to draw a large clear diagram showing all forces acting.

Resolving horizontally:
$$\rightarrow T_1 \cos(60° + \theta) - T_2 \cos(60° - \theta) = 0$$

Since $T_1 = 2T_2$ and using the result for $\cos(A + B)$

See Book P2 page 62.

$$2T_2(\cos 60° \cos \theta - \sin 60° \sin \theta) = T_2(\cos 60° \cos \theta + \sin 60° \sin \theta)$$

$$2\left(\tfrac{1}{2}\cos \theta - \frac{\sqrt{3}}{2}\sin \theta\right) = \tfrac{1}{2}\cos \theta + \frac{\sqrt{3}}{2}\sin \theta$$

or
$$\tfrac{1}{2}\cos \theta = \frac{3\sqrt{3}}{2}\sin \theta$$

So
$$\tan \theta = \frac{1}{3\sqrt{3}} \text{ as required.}$$

(a) Resolving vertically:
$$\uparrow T_1 \sin(60° + \theta) + T_2 \sin(60° - \theta) - (M + m)g = 0$$

Using $T_1 = 2T_2$ and the result for $\sin(A + B)$

See Book P2 page 62.

$$2T_2(\sin 60° \cos \theta + \cos 60° \sin \theta) + T_2(\sin 60° \cos \theta - \cos 60° \sin \theta)$$
$$-(M + m)g = 0$$

So:
$$T_2\left(2\frac{\sqrt{3}}{2}\cos \theta + 2\left(\tfrac{1}{2}\right)\sin \theta + \frac{\sqrt{3}}{2}\cos \theta - \tfrac{1}{2}\sin \theta\right) = (M + m)g$$

$$T_2\left(\frac{3\sqrt{3}}{2}\cos \theta + \tfrac{1}{2}\sin \theta\right) = (M + m)g$$

But $\tan \theta = \dfrac{1}{3\sqrt{3}}$

so $\sin \theta = \dfrac{1}{\sqrt{28}}$

$\cos \theta = \dfrac{3\sqrt{3}}{\sqrt{28}}$

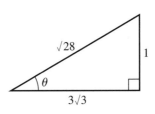

So:
$$T_2\left(\frac{3\sqrt{3}}{2}\frac{3\sqrt{3}}{\sqrt{28}} + \tfrac{1}{2}\frac{1}{\sqrt{28}}\right) = (M + m)g$$

$$T_2\left(\frac{28}{2\sqrt{28}}\right) = (M + m)g$$

$$\frac{T_2}{2}\sqrt{28} = (M + m)g$$

$$T_2\sqrt{7} = (M + m)g$$

So:
$$T_2 = \frac{1}{\sqrt{7}}(M + m)g$$

and
$$T_1 = 2T_2 = \frac{2}{\sqrt{7}}(M + m)g$$

(b) The vertical intersects $CDEF$ at Q.

Taking moments about A:

> The point A is chosen as both T_1 and T_2 act through A.

$$\curvearrowright MgQN\cos\theta - mgQP\cos\theta = 0$$

But $\dfrac{QN}{NA} = \tan\theta$

So $QN = a\sqrt{3}\tan\theta$

Also $QP = NP - QN$

$$= (3a - x) - a\sqrt{3}\tan\theta$$

So: $M \times QN - m \times QP = Ma\sqrt{3}\tan\theta - m(3a - x - a\sqrt{3}\tan\theta) = 0$

Using $\tan\theta = \dfrac{1}{3\sqrt{3}}$

$$Ma\frac{\sqrt{3}}{3\sqrt{3}} - 3am + mx + \frac{ma\sqrt{3}}{3\sqrt{3}} = 0$$

$$-\frac{Ma}{3} + 3am - \frac{ma}{3} = mx$$

$$x = \frac{1}{m}\left(\frac{8ma}{3} - \frac{Ma}{3}\right)$$

$$= \frac{8a}{3} - \frac{M}{m}\frac{a}{3}$$

$$= \frac{a}{3}\left(8 - \frac{M}{m}\right)$$

Revision exercise 5

1

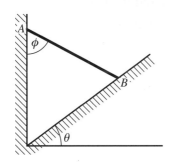

The diagram shows a uniform ladder AB resting in equilibrium with end A in contact with a smooth vertical wall and end B in contact with a smooth inclined plane which makes an angle θ with the horizontal. Given that the ladder makes an angle ϕ with the vertical, show that $\tan\phi = 2\tan\theta$.

2 A uniform ladder of length $2a$ rests in limiting equilibrium in a vertical plane with its lower end on rough horizontal ground and its upper end against a smooth vertical wall. The ladder makes an angle $60°$ with the ground. Show that the coefficient of friction is $\sqrt{3}/6$.

The ladder is lowered in its vertical plane, whilst still resting against the smooth wall and the ground, to make an angle $30°$ with the ground. The coefficient of friction between the ladder and the ground remains at $\sqrt{3}/6$. A man whose weight is four times that of the ladder starts climbing up the ladder. Find how far he can climb up the ladder before it slips.   [E]

3 A uniform plank AB, of length 3 m and mass 5 kg, is freely hinged at A to a vertical wall. The plank is held horizontally, in equilibrium, by an inextensible string, one end of which is attached to B and the other to the point C on the wall, where C is 3 m above A. Find the magnitude of the reaction at A.

4

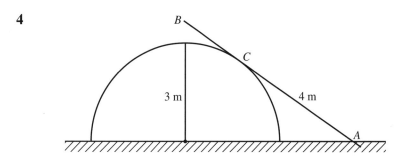

The diagram shows a smooth hemisphere, of radius 3 m, fixed with its plane face on a rough horizontal floor. A uniform ladder AB, of length 5 m and mass 12 kg, rests in equilibrium in a vertical plane with end A on the floor and a point C of the ladder in contact with the hemisphere, where $AC = 4$ m.

Given that the ladder is on the point of slipping find,
(a) the reaction at C,
(b) the coefficient of friction between the ladder and the floor.

Test yourself	**What to review**
	If your answer is incorrect:
1 A straight uniform rod AB, of mass M, rests in equilibrium with the end A in contact with horizontal ground and the end B against a smooth vertical wall. The vertical plane containing AB is at right angles to the wall and AB is inclined at $60°$ to the horizontal. The coefficient of friction between the rod and the ground is μ. **(a)** Find, in terms of M and g, the magnitude of the force exerted by the wall on the rod. **(b)** Show that $\mu \geqslant \frac{1}{6}\sqrt{3}$. A load of mass kM is attached to the rod at B, where k is a positive constant. **(c)** Given that $\mu = \frac{1}{5}\sqrt{3}$ and that equilibrium is limiting, find the value of k. [E]	*Review Heinemann Book M2 pages 136–141*
2 A uniform rod AB, of mass m and length l, is smoothly jointed at the end A to a fixed straight horizontal wire AC. The end B is attached by means of a light inextensible string, also of length l, to a small ring, of mass m, which can slide on the wire, the coefficient of friction between the ring and the wire being μ. The rod is in equilibrium in the vertical plane through the wire. Given that α is the inclination of the string to the horizontal, show that the tension in the string is of magnitude $\dfrac{mg}{(4\sin\alpha)}$. Show also that $\tan\alpha \geqslant \dfrac{1}{5\mu}$ [E]	*Review Heinemann Book M2 pages 129–133*

Test yourself answers

1 (a) $\frac{1}{6}Mg\sqrt{3}$ (c) $k = \frac{1}{4}$

Examination style paper

Answer all questions   **Time allowed 90 minutes**

Whenever a numerical value of g is required, take $g = 9.8\,\mathrm{m\,s^{-2}}$.

1. A vertical tower stands on level ground. A stone is thrown from the top of the tower with an initial speed of $23.5\,\mathrm{m\,s^{-1}}$ at $\arctan\left(\frac{4}{3}\right)$ above the horizontal. The stone strikes the ground at a point $70.5\,\mathrm{m}$ from the foot of the tower. Find:
   (a) the time taken for the stone to reach the ground,   **(4 marks)**
   (b) the height of the tower.   **(4 marks)**

2. A car of mass $1000\,\mathrm{kg}$ is moving at a constant speed of $25\,\mathrm{m\,s^{-1}}$ up a hill inclined at $\arcsin\left(\frac{1}{14}\right)$ to the horizontal. The power developed by the car's engine is $24\,\mathrm{kW}$.
   (a) Find the resistance to the motion.   **(7 marks)**
   At the top of the hill the road becomes horizontal.
   (b) Assuming the resistance to be unchanged, find the initial acceleration of the car.   **(3 marks)**

3.

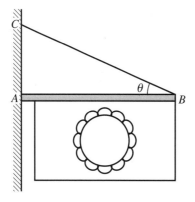

The figure shows a pole of mass m and length l displaying a light banner. The pole is modelled as a uniform rod AB, freely hinged to a vertical wall at the point A. It is held in a horizontal position by a light wire inclined at angle θ to the horizontal. One end of this wire is attached to the end B of the rod and the other end is attached to the wall at the point C which is vertically above A and such that $\tan\theta = \frac{1}{3}$.
   (a) Show that the tension in the wire is $\frac{1}{2}\sqrt{10}\,mg$.   **(3 marks)**
   (b) Find the magnitude of the force exerted by the wall on the rod at A.   **(7 marks)**

4. At time $t = 0$ a particle P is at the point with position vector $(10\mathbf{i} + 7\mathbf{j})$ m relative to an origin O. The particle moves with constant acceleration $\ddot{\mathbf{r}}$ where $\ddot{\mathbf{r}} = (4\mathbf{i} + 3\mathbf{j})$ m s^{-2}. Given that when $t = 0$, $\dot{\mathbf{r}} = \mathbf{0}$ find
    (a) the velocity, $\dot{\mathbf{r}}$, when $t = 4$ seconds, **(4 marks)**
    (b) the distance of P from O at this time. **(7 marks)**

5.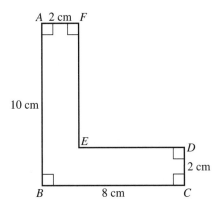

    The diagram shows a figure made from a uniform sheet of metal. Find the distance, in cm, of the centre of mass of the shape
    (a) (i) from AB,
         (ii) from BC. **(8 marks)**
    The shape hangs in equilibrium from the point A.
    (b) Find, to the nearest degree, the angle made by AB with the downward vertical. **(3 marks)**

6. A particle P of mass 2 kg is projected up a rough plane, inclined at $\arcsin\left(\frac{5}{13}\right)$ to the horizontal, with a speed of 5 m s^{-1}. Given that the coefficient of friction between the particle and the plane is $\frac{1}{2}$, find the distance, in metres, travelled by P before coming to rest. **(12 marks)**

7. A small smooth sphere of mass m moving on a horizontal plane with a speed V collides directly with a small smooth sphere of mass $2m$ at rest. The coefficient of restitution between the two spheres is e.
    (a) Find the speeds of the two spheres after the impact. **(7 marks)**
    Given that one half of the initial kinetic energy is lost in the impact,
    (b) find the value of e. **(6 marks)**

Answers

Revision exercise 1A

1 $8\frac{4}{7}$ or 8.57 m

2 **(b)** 0.8

3 **(a)** horizontal velocity same for aircraft and bomb
  (b) $14\frac{2}{7}$ s or 14.29 s **(c)** 1400 m
  (d) 140 m s^{-1}

4 **(a)** 3.4 **(b)** 38 **(c)** 14 **(d)** 12 **(e)** 15°

5 **(a)** $y = x - \dfrac{x^2}{49\,000}$ **(b)** 49 000 m **(c)** 50 s
  (d) 490 m s^{-1}

6 **(a)** 35 **(b)** 31

7 **(b)** 40 m **(c)** 221 m
  (d) 45 m s^{-1}, 10.7° below horizontal

8 **(a)** 8.1 s **(c)** 14.7 m s^{-1} **(d)** 35.8, 33.2

Revision exercise 1B

1 **(a)** 2 s, 6 s

2 **(a)** 3 **(b)** 18 m

3 **(a)** 2 s, 4 s **(b)** 13 m

4 **(a)** 37 m s^{-1} **(b)** $t^3 + t^2 + 4t$ **(c)** 76 m
  (d)

5 **(a)** 10 m s^{-2} **(b)** $\{(6t-2)\mathbf{i} + (8t-5)\mathbf{j}\}$ m s^{-1}
  (c) 1.5 s

6 **(a)** 15 m s^{-1}
  (b) $2\sqrt{2}$ m s^{-1} or 2.83 m s^{-1}, 135°

7 **(a)** 14 m s^{-2} **(b)** $70\frac{2}{3}$ m **(c)** $(6\mathbf{i} - 8\mathbf{j})$ m s^{-1}

(d) $(\mathbf{i} + 8\mathbf{j})$ m s^{-1} **(e)** $\sqrt{65}$ m s^{-1} or 8.06 m s^{-1}
(f) 82.9°

Revision exercise 2

1 (5, 0)

2 **(a)** $1\frac{3}{4}$ cm **(b)** $1\frac{11}{13}$ cm

3 **(b)** 34°

4 $\dfrac{4}{3\pi}$

5 **(b)** 95.7°

6 **(a)** $\dfrac{6l}{5}$ **(b)** l **(c)** 51°

7 **(a)** $\dfrac{8l}{7}$ **(b)** $\dfrac{13l}{7}$ **(c)** $\frac{1}{8}$

Revision exercise 3

1 **(a)** 2940 J **(b)** 65.3 W

2 6220 J

3 **(a)** 78.4 N **(b)** 114 N **(c)** 7.16 m s^{-1}

4 **(a)** 40 kW **(b)** 48 kW **(c)** 12.7 m s^{-1}

5 **(a)** 320 kW **(b)** 0.076 m s^{-2}

6 **(a)** 960 N **(b)** 0.75 **(c)** 360 N

7 **(a)** 30 m s^{-1} **(b)** 2600 N **(c)** 19 m s^{-1}
  (d) 38.8 m

8 **(a)** $K = \dfrac{R}{100} + \dfrac{2}{3}g$ **(b)** 523 **(c)** 11.8

Revision exercise 4

1 **(a)** u, $2u$ **(b)** $3mu$

2 **(a)** 1.9 m s^{-1} **(b)** $\frac{1}{2}$

3 **(a)** $v = 8u$, speed of $A = u$ **(b)** $9mu$
  (c) $18mu^2$

4 $\dfrac{\pi a}{2T}(1 - e)$, $\dfrac{\pi a}{2T}(1 + e)$

6 $\frac{5}{8}$

Revision exercise 5

2 $\frac{1}{6}a$

3 34.6 N

4 **(a)** 58.8 N **(b)** $\frac{1}{2}$

Examination style paper

1 **(a)** 5 s **(b)** 28.5 m

2 **(a)** 260 N **(b)** $0.7 \, \text{m s}^{-2}$

3 **(b)** $\frac{1}{2}mg\sqrt{10}$

4 **(a)** $(16\mathbf{i} + 12\mathbf{j}) \, \text{m s}^{-1}$

    (b) 52.2 m

5 **(a)** (i) 2.5 cm (ii) 3.5 cm

    (b) 21°

6 1.51 m

7 **(a)** $\frac{1}{3}(1 - 2e)V, \frac{1}{3}(1 + e)V$

    (b) $e = \frac{1}{2}$

Heinemann Modular Mathematics for Edexcel AS and A Level

The leading course for Edexcel A Level

Heinemann Modular Mathematics for **Edexcel AS and A Level** provides dedicated syllabus-specific textbooks to give you a clear route to success. The clear layout and structure of the books match the flexible modular structure of the course, meaning that you only study the elements you need to. Perfect for use in the classroom or at home, the series covers the topics at the level and depth that you'll need. Each book includes lots of test questions and answers, and features a full mock-exam paper.

All the books have been written by experienced teachers who have been involved with the development of the new specification, so you can be sure that these are the right textbooks for you.

To see any of the following titles FREE for 60 days or to order your books straight away call Customer Services on 01865 888068

Pure Mathematics 1 (P1)
0434 510886

Pure Mathematics 2 (P2)
0435 510894

Pure Mathematics 3 (P3)
0435 510908

Pure Mathematics 4 (P4)
0435 510916

Pure Mathematics 5 (P5)
0435 510924

Pure Mathematics 6 (P6)
0435 510932

Mechanics 1 (M1)
0435 510746

Mechanics 2 (M2)
0435 510754

Mechanics 3 (M3)
0434 510762

Mechanics 4 (M4)
0435 510770

Mechanics 5 (M5)
0435 510789

Mechanics 6 (M6)
0435 510797

Statistics 1 (S1)
0435 510827

Statistics 2 (S2)
0435 510835

Statistics 3 (S3)
0435 510843

Statistics 4 (S4)
0435 510851

Statistics 5 (S5)
0435 51086X

Statistics 6 (S6)
0435 510878

Decision Mathematics (D1)
0435 510800

Decision Mathematics (D2)
0435 510819

S 999 ADV 08